China and the Internet

China and the Internet: Politics of the digital leap forward is a comprehensive assessment of the political and economic impact of information and communication technologies (ICTs) on Chinese society. It provides in-depth analyses of topics including economic development, civil and political liberties, bureaucratic politics, international relations and security studies.

The book covers the aspirations of Chinese policy-makers for using the Internet to achieve a 'digital leapfrog' of economic development. After looking at the achievements made so far putting China online, the authors consider prospects for realising this digital leapfrog from a number of perspectives, including the nature of the one-party state, the lack of visible legal structure, the digital divide within China and the problem of bureaucratic turf wars and the technological time lag behind the advanced industrial economies. The chapters also explore normative issues, such as the state's ability to maintain control over the flow of information and to stifle political dissent, the impact of the Internet on the formation of national and regional identity and the impact of ICTs on international security.

Throughout, the authors explore methodological and theoretical problems – from the reliability of data to the nature of the relationship between technological and social change. Avoiding technical jargon, the book is accessible to anyone interested in the social impact of the Internet and information and communication technologies, from academia to business and public policy-making.

Christopher R. Hughes is Director of the Asia Research Centre and Lecturer in the Department of International Relations at the London School of Economics. **Gudrun Wacker** is Head of the Research Unit Asia, Stiftung Wissenschaft und Politik (German Institute for International and Security Affairs), Berlin, Germany's major think-tank.

Politics in Asia series
Formerly edited by Michael Leifer
London School of Economics

ASEAN and the Security of South-East Asia
Michael Leifer

China's Policy towards Territorial Disputes
The case of the South China Sea Islands
Chi-kin Lo

India and Southeast Asia
Indian perceptions and policies
Mohammed Ayoob

Gorbachev and Southeast Asia
Leszek Buszynski

Indonesian Politics under Suharto
Order, development and pressure for change
Michael R.J. Vatikiotis

The State and Ethnic Politics in Southeast Asia
David Brown

The Politics of Nation Building and Citizenship in Singapore
Michael Hill and Lian Kwen Fee

Politics in Indonesia
Democracy, Islam and the ideology of tolerance
Douglas E. Ramage

Communitarian Ideology and Democracy in Singapore
Beng-Huat Chua

The Challenge of Democracy in Nepal
Louise Brown

Japan's Asia Policy
Wolf Mendl

The International Politics of the Asia-Pacific, 1945–1995
Michael Yahuda

Political Change in Southeast Asia
Trimming the banyan tree
Michael R.J. Vatikiotis

Hong Kong
China's challenge
Michael Yahuda

Korea versus Korea
A case of contested legitimacy
B.K. Gills

Taiwan and Chinese Nationalism
National identity and status in international society
Christopher Hughes

Managing Political Change in Singapore
The elected presidency
Kevin Y.L. Tan and Lam Peng Er

Islam in Malaysian Foreign Policy
Shanti Nair

Political Change in Thailand
Democracy and participation
Kevin Hewison

The Politics of NGOs in South-East Asia
Participation and protest in the Philippines
Gerard Clarke

Malaysian Politics Under Mahathir
R.S. Milne and Diane K. Mauzy

Indonesia and China
The politics of a troubled relationship
Rizal Sukma

Arming the Two Koreas
State, capital and military power
Taik-young Hamm

Engaging China
The management of an emerging power
Edited by Alastair Iain Johnston and Robert S. Ross

Singapore's Foreign Policy
Coping with vulnerability
Michael Leifer

Philippine Politics and Society in the Twentieth Century
Colonial legacies, post-colonial trajectories
Eva-Lotta E. Hedman and John T. Sidel

Constructing a Security Community in Southeast Asia
ASEAN and the problem of regional order
Amitav Acharya

Monarchy in South-East Asia
The faces of tradition in transition
Roger Kershaw

Korea After the Crash
The politics of economic recovery
Brian Bridges

The Future of North Korea
Edited by Tsuneo Akaha

The International Relations of Japan and South East Asia
Forging a new regionalism
Sueo Sudo

Power and Change in Central Asia
Edited by Sally N. Cummings

The Politics of Human Rights in Southeast Asia
Philip Eldridge

Political Business in East Asia
Edited by Edmund Terence Gomez

Singapore Politics under the People's Action Party
Diane K. Mauzy and R.S. Milne

Media and Politics in Pacific Asia
Duncan McCargo

Japanese Governance
Beyond Japan Inc
Edited by Jennifer Amyx and Peter Drysdale

China and the Internet
Politics of the digital leap forward
Edited by Christopher R. Hughes and Gudrun Wacker

China and the Internet
Politics of the digital leap forward

Edited by Christopher R. Hughes
and Gudrun Wacker

LONDON AND NEW YORK

To the memory of Michael Leifer

First published 2003
by RoutledgeCurzon
11 New Fetter Lane, London EC4P 4EE

Simultaneously published in the USA and Canada
by RoutledgeCurzon
29 West 35th Street, New York, NY 10001

RoutledgeCurzon is an imprint of the Taylor & Francis Group

© 2003 Christopher R. Hughes and Gudrun Wacker for selection and editorial matter; individual chapters the contributors

Typeset in Sabon by
Keystroke, Jacaranda Lodge, Wolverhampton
Printed and bound in Great Britain by
MPG Books Ltd, Bodmin

All rights reserved. No part of this book may be reprinted or reproduced or utilised in any form or by any electronic, mechanical, or other means, now known or hereafter invented, including photocopying and recording, or in any information storage or retrieval system, without permission in writing from the publishers.

British Library Cataloguing in Publication Data
A catalogue record for this book is available from the British Library

Library of Congress Cataloging in Publication Data
China and the Internet: politics and the digital leap forward/edited by Christopher R. Hughes and Gudrun Wacker.
  p. cm. – (Politics in Asia series)
"The papers collected together in this volume . . . are the product of two workshops organised between the Asia Research Centre of the London School of Economics and the Stiftung Wissenschaft und Politik (SWP), Berlin, on 8–9 December 2000 and 21–23 February 2002"–P.
Includes bibliographical references and index.
1. Information technology–Economic aspects–China–Congresses.
2. Information technology–Political aspects–China–Congresses.
3. Internet–Economic aspects–China–Congresses.
4. Internet–Political aspects–China–Congresses. I. Hughes, Christopher R., 1960– II. Wacker, Gudrun, 1954– III. Series.

HC430.I55 C45 2003
303.48′33′0951–dc21   2002033367

ISBN 0–415–27772–8

Contents

List of illustrations vii
Notes on contributors ix
List of abbreviations xi

Introduction: China's digital leap forward 1
CHRISTOPHER R. HUGHES AND GUDRUN WACKER

1 ICTs in China's development strategy 8
  XIUDIAN DAI

2 Internet growth and the digital divide: implications for spatial development 30
  KARSTEN GIESE

3 The Internet and censorship in China 58
  GUDRUN WACKER

4 Network convergence and bureaucratic turf wars 83
  JUNHUA ZHANG

5 (Re-)Imagining 'Greater China': Silicon Valley and the strategy of siliconization 102
  NGAI-LING SUM

6 What's in a name? China and the Domain Name System 127
  MONIKA ERMERT AND CHRISTOPHER R. HUGHES

7 Fighting the smokeless war: ICTs and international security 139
  CHRISTOPHER R. HUGHES

Bibliography 162
Index 175

Illustrations

Figures

1.1	Growth of total Internet bandwidth for international traffic (1997–2002)	16
2.1	Internet growth in China (1997–2002)	31
2.2	Internet hosts in Mainland China (1997–2002)	34
2.3	Chinese Internet users and WorldWideWeb contents (1997–2002)	36
2.4	Second level domains under '.cn' (1997–2002)	36
2.5	Geographic concentration of Internet users, Websites and domains (2002)	42
2.6	Prevalence of netizens in selected regions (1997 and 2002)	44

Maps

2.1	Density of rural telephones and urban mobile phone subscribers in China (2000)	39
2.2	Wealth distribution and urban computer ownership in China	41
2.3	Concentration of WorldWideWeb users in China (2001)	43
2.4	Illiteracy rates and government funding for education (2000)	45
2.5	Individual incomes and government support for underdeveloped regions (2000)	47
2.6	Urbanisation, foreign direct investment and Internet penetration rates	49
2.7	UNDP Internet projects in China	51

Tables

1.1	The costs of communications	11
1.2	China's information superhighways: fibre-optic cable links (1991–95)	14
1.3	Internet backbone networks and their international bandwidth (January 2002)	16

2.1	CNNIC surveys on Internet development in China	32
4.1	Broadband specifications of seven licensed state-owned network carriers	86
4.2	International comparison of household penetration of networks (1998)	91
4.3	Increase of cable TV subscriber numbers	91
5.1	A new economic object of 'growth': Cyberport	112
5.2	Nankang Software Park	115
5.3	CyberCity Shenzhen	119

Contributors

Xiudian Dai is Senior Lecturer in the Department of Politics and International Studies at the University of Hull, specialising in the political economy of new media technologies with particular reference to Europe and East Asia. He is the author of *Corporate Strategy, Public Policy and New Technologies* (Pergamon 1996) and *The Digital Revolution and Governance* (Ashgate 2000).

Monika Ermert is a freelance journalist who has published widely on issues concerning the politics of the Internet, ICANN and the Domain Name System, with special reference to China, in leading organs such as *Die Zeit* and *Suddeutsche Zeitung*.

Karsten Giese is a Research Fellow at the Institute of Asian Affairs, Hamburg. His research interests include social and cultural change, Internet development and migration in China, as well as Chinese foreign and security policy. He has published articles on the Internet and political economy in China in various books and journals, including *China aktuell*.

Christopher R. Hughes is Director of the Asia Research Centre and Lecturer in the Department of International Relations at the London School of Economics, with special reference to China and the Asia Pacific. He has written widely on Chinese international relations and published articles on information and communication technologies in China in *The Cambridge Review of International Affairs, New Media and Society* and *The World Today*.

Ngai-Ling Sum is Lecturer in the Department of Politics at the University of Lancaster. She has research interests in globalisation, transborder regions and the political economy of the information age. She is on the editorial board of *The Pacific Review*. Her most recent publications include *Globalization, Regionalization and Cross-Border Regions* (edited with Markus Perkmann) and articles in *Critical Asian Studies, Capital & Class, New Political Economy, Urban Studies* and *Economy and Society*, as well as contributions to various edited volumes.

Gudrun Wacker is head of the Asia research unit at the Stiftung Wissenschaft und Politik (German Institute for International and Security Affairs), Berlin, Germany's major think-tank. Her research is focused on Chinese foreign and security policy and she has published widely on these topics.

Junhua Zhang is a Lecturer at the Free University of Berlin with special research interests in Chinese and East Asian politics. He has published articles on information and communication technologies in *Asien* and *New Media and Society*.

Abbreviations

3G	Third generation mobile communications
ABA	Administration of Broadcasting Affairs
ADSL	Asynchronous digital subscriber line
APEC	Asia Pacific Economic Cooperation
ASCII	American Standard Code for Information Interchange
AWACS	Airborne warning and control systems
B2B	Business-to-business (e-commerce)
B2C	Business-to-consumer (e-commerce)
BBS	Bulletin board service
C3I	Command, control, communications and intelligence
CAINET	China Advanced Info-Optical Network
CANET	Chinese Academic Network
CAS	Chinese Academy of Sciences
CATV	Cable television
CCP	Chinese Communist Party
ccTLD	Country code Top Level Domain
CCTV	China Central Television
CDMA	Code division multiple access
CEO	Chief executive officer
CERNET	China Education and Research Network
CMC	China Mobile Communications
CNIX	China Internet Exchange
CNJ	China Netcom Jitong
CNNIC	China Internet Network Information Center (*Zhongguo hulian wangluo xinxi zhongxin*)
CPPC	Chinese People's Political Consultative Conference
CPU	Central Processing Unit
CR	China Railcom
CRFTG	China Radio, Film and Television Group
CSC	China Satellite Communications Group
CSTNET	China Science and Technology Network
CT	China Telecom
CU	China Unicom
DNS	Domain Name System

DPP	Democratic Progressive Party of Taiwan
FCC	Federal Communications Commission (US)
FDI	Foreign direct investment
GAC	General Advisory Committee (ICANN)
GDP	Gross domestic product
GEM	Growth Enterprise Market
GII	Global Information Infrastructure
GSM	Global standard for mobile communications
gTLD	Generic Top Level Domains
HFC	Hybrid Fibre Coaxial
IANA	Internet Assigned Numbers Authority
ICA	Institute of Computer Application (Beijing)
ICANN	Internet Corporation for Assigned Names and Numbers
ICP	Internet content provider
ICTs	Information and communications technologies
IETF	Internet Engineering Task Force
IIF	Initial installation fee
INE	Informatization of the national economy
IP	Internet protocol
IPO	Initial public offering
IPRs	Intellectual property rights
IP-VPNs	Internet protocol virtual private networks
ISDN	Integrated services digital network
ISO	International Standards Organisation
ISP	Internet service provider
IT	Information technology
ITC	Independent Television Commission (UK)
ITDLG	Information Technology Development Leading Group
ITU	International Telecommunications Union
JCINE	Joint Conference for the Informatization of the National Economy
KISA	Korea Information Security Agency
LSIC	Large-scale integrated circuits
MEI	Ministry of Electronic Industry
MII	Ministry of Information Industry
MINC	Multilingual Internet Names Consortium
MPT	Ministry of Posts and Telecommunications
MRFT	Ministry of Radio, Film and Television
NAP	Network access point
NATO	North Atlantic Treaty Organisation
NII	National Information Infrastructure
NSI	Network Solutions Inc.
NTD	New Taiwan Dollar
NTIA	National Telecommunications and Information Association (US)

OEM	Original equipment manufacturing
Ofcom	Office for Communications (UK)
Oftel	Office for Telecommunications (UK)
ORSC	Original equipment manufacturing
PICS	Platform for Internet content selection
PLA	People's Liberation Army
PRC	People's Republic of China
RFC	Request for Comment
RIP	Regulation of Investigatory Powers (UK Act)
RMA	Revolution in Military Affairs
RMB	*Renminbi Yuan*
SARFT	State Administration of Radio, Film and Television
SCIMC	State Council Information Management Commission
SDPC	State Development Planning Commission
SEZ	Special economic zone
SHKP	Sun Hung Kai Properties
SMEs	Small and medium-sized enterprises
SWP	Stiftung Wissenschaft und Politik (Berlin)
TD-SCDMA	Time division-synchronous code division multiple access
TLD	Top Level Domain
UNDP	United Nations Development Programme
URL	Universal Resource Locator
USD	United States Dollar
VoD	Video on demand
VoIP	Voice over IP
W-CDMA	Wide-code division multiple access
WAP	Wireless Application Protocol
WTO	World Trade Organisation

Introduction
China's digital leap forward

Christopher R. Hughes and Gudrun Wacker

The study of Chinese politics can no longer be considered complete without an understanding of the social impact of information and communication technologies (ICTs). Since the country's first e-mail was sent overseas in September 1987, the number of Internet users has risen to over 30 million, a Ministry of Information Industry (MII) has been established, ICTs have become a central element of the Five-Year Plan, and there is a wide-ranging debate over their impact across the spectrum of public policy. This volume aims to go some way towards clarifying the relationship between technology and politics in China that is indicated by such developments. It is the product of two workshops organised between the Asia Research Centre of the London School of Economics and the Stiftung Wissenschaft und Politik (SWP), Berlin, on 8–9 December 2000, and 21–23 February 2002, generously supported by the Fritz Thyssen Foundation and the British Academy respectively.

While the original task of the participants in this project was to focus on the social and political impact of the Internet in China, it quickly became clear that convergence between different kinds of information technologies meant that the picture had to be somewhat broader, touching at least on telephone, cable television and satellite systems as well. The Internet, however, remains the main case study for the book, and the focus has been fixed on politics by locating its appropriation within the context of public policy making. While it has been necessary to use some technical language and explanations in the process, the authors have tried to keep this to a minimum so that their work will be accessible to as broad a range of social scientists as possible.

There are two main reasons for the title of the book, *China and the Internet: Politics of the digital leap forward*. First, it is an allusion to the belief amongst China's political leaders that a developmental 'leapfrog' can be achieved on the back of ICTs. Second, it is intended as a reminder that this is not the first attempt to make a developmental 'leap forward' in China's modern history. In other words, the title is also supposed to emphasise the view of the authors that the political impact of new technologies needs to be understood in the context of modern Chinese history. While the belief in a 'digital leapfrog' is a long way from the political radicalism of Mao Zedong's 'Great Leap Forward', both movements reflect a deep-seated tendency to try to catch up with the advanced industrialised world through unorthodox means. This is, of course, symptomatic of a deep-seated Promethean obsession with science and technology that Mao and his successors as leaders of the Chinese Communist Party (CCP) took from an intellectual tradition that goes back all the way to the neo-Confucian reformers who advocated using Western functional knowledge (*yong*) to preserve Chinese cultural essence (*ti*) when the Qing dynasty was faced by destruction at the hands of Western imperialism.[1] This remains the sentiment that lies behind the campaign of the 'Three Representations' launched by CCP General Secretary Jiang Zemin in the late 1990s, according to which the party should represent 'China's advanced productive forces' and 'China's advanced culture'.

It is within this political culture that ICTs have risen to become an important area not only of policy-making, but of the nation-building legitimacy of the Chinese leadership too. In this volume, Xiudian Dai (Chapter 1) demonstrates the strength of the belief among China's leaders that ICTs can play a major role in bringing about the economic growth upon which the CCP has pinned its legitimacy since the late 1970s in his chapter on ICTs in China's development strategy. He also stresses the way in which this was strongly influenced by the global climate of opinion in the 1990s, when influential international organisations such as the European Commission, the United Nations, the G7/8 and the World Economic Forum, encouraged China's leaders to believe that cheaper and more efficient communications would lead to lower costs and greater prospects for economic growth. Policy-makers have found such a vision so attractive that they have even come to believe that economic underdevelopment can be an asset when it comes to developmental 'leapfrogging', because obstacles like the presence of a well-established copper wire system do not stand in the way of going straight for the use of cutting-edge technologies, like fibre optics. The consequent achievements that have been made in terms of building a massive national information infrastructure are indeed quite remarkable in terms of kilometres of cable, the rapidly rising number of 'netizens', and the proliferation of Chinese-language websites. There is even reported to be life in the e-commerce sector, with farmers said to be using the Internet to get connected to domestic and international markets.

Yet, as Dai also argues, such successes have also brought problems, most of which are generated by the nature of the political system in China. The need for the state bureaucracy to keep up with the pace of technological change, for example, necessitates a large degree of restructuring. Equally problematic is the still weak respect for the rule of law in the People's Republic of China (PRC), with favoured constituencies finding it easy to create anomalies when required to conform to the rules issued by central regulators. The ultimate dilemma, though, is that of a ruling party that wants the benefits of informatisation, while lacking the will to promote active participation by citizens or to relax its control over the provision of content.

Karsten Giese (Chapter 2) draws attention to the limitations imposed on the economic potential of the Internet by China's social conditions in his chapter on the digital divide. Reminding us that the data upon which much of the talk about the rapid development of the Internet in China is based comes from notoriously unreliable statistics, he goes on to explain how advocates of the digital leapfrog face the remarkably traditional hurdle of underdevelopment in large parts of the country. Thus, while global trends indicate that the country's urban market for information services may already be near saturation point, expanding into new domestic markets means overcoming serious problems of illiteracy, poor education and training, lack of basic infrastructure and a paucity of investment. Yet while the state is unwilling to take a decisive lead in remedying such deficiencies, it cannot rely wholly on attracting the private sector to invest in areas where the conditions for high-tech industries are far from viable. If such a view is correct, then encouraging informatisation for the sake of rapid development may actually make the digital divide inside China even wider.

Aside from the issue of the impact of ICTs on economic development, students of Chinese politics also have to deal with the question of whether technologies like the Internet undermine the power of authoritarian regimes. There has, of course, been much speculation in popular debates around the world that such a trend is inevitable. Underpinning such a belief seems to be an irresistibly simple formula: states of all kinds have to encourage the spread of ICTs if they want to enjoy the global economic boom, but by enhancing the free flow of information within their own societies and across borders they will inevitably undermine their own grip on political power. As the information revolution really took off just as the Cold War was coming to an end, it is not hard to see how such a thesis could be encouraged by the tide of democratisation that accompanied the fall of communist regimes around the world. The case of the People's Republic of China, however, seems to pose something of an embarrassing exception. Ten years have already passed since the National Science Foundation of the United States lifted the ban on commercial activity on the Internet in 1992. China was quick to get connected to the World Wide Web after its invention in 1991, and to open its doors to the new flood of customer-based technologies that have allowed the Internet to become a popular means of communication. Yet as research on the social and political impact of this explosion has reached new levels of sophistication, there seems to be little evidence to support the thesis that the Chinese state is being eroded in any special way by ICTs. On the contrary, recent research seems to suggest that states around the world are having a degree of success in staging something of a counter-revolution.[2]

Gudrun Wacker (Chapter 3) explores the relationship between the political stability of the authoritarian state and the Internet by adopting a theoretical framework that draws on the ideas of Lawrence Lessig, the world's foremost writer on the impact of the Internet on civil liberties. She also uses the ideas of James Boyle, who has applied a Foucauldian analysis to understand how control is made possible by inculcating the widespread belief that the state just might be watching individual users. The analogy can thus be drawn with Jeremy Bentham's 'Panopticon', the prison in which the interiors of all the cells are fully exposed to a central warden. Because the warden's actions are invisible to the inmates, and they are thus unable to know when he is watching them, the actions of an almost unlimited number of prisoners can be controlled by just one person. The role of the warden in China's Internet panopticon is played by a combination of state security agencies working with commercial actors, both domestic and foreign, who supply the necessary technology and expertise to present the credible possibility that the actions of users just might be under surveillance. If this is correct, then the Internet is unlikely to be forum within which democratic dissent is expressed and organised, although it may well act as a catalyst for change if instability arises for other reasons.

Wacker's argument should thus serve as something of a rude awakening from what seems to have been a severe case of technological determinism for those who are wedded to the belief that the social impact of technology can be grasped in the absence of its political context. This would be of little surprise, perhaps, to those more familiar with the sociology of technology. In this sub-field of the social sciences students are well aware of the ways in which technologies are not 'neutral' but have a social impact that is determined by the purpose to which they are put. One of the best known illustrations of this is the story of Robert Moses, the builder who constructed bridges in New York from the 1920s to the 1970s with such a low height that buses carrying the poor would not be able to pass beneath them and reach the beaches of Long Island.[3] Just as a bridge can be used to divide rather than connect, so information technology can be used for a variety of purposes that may have little to do with genuine communication.

Yet the sociology of technology also teaches us not to assume that there are no limits at all to the ways in which technologies can be manipulated to promote particular political agendas. As Langdon Winner has argued, there are cases where some technologies seem to be 'inherently political' in the ways that they impose certain kinds of structures and practices on societies. Nuclear power, for example, seems to demand some sacrifice of civil liberties and the imposition of certain authoritarian structures if it is to be exploited without the risk of terrorists and other criminals getting their hands on materials essential for the production of weapons of mass destruction.[4] In this volume, the chapters by Junhua Zhang, Ngai-Ling Sum and Monika Ermert and Christopher R. Hughes all illustrate how certain characteristics of ICTs may also make them inherently political technologies in the same way.

Central to Zhang's analysis (Chapter 4) is an exploration of the way in which convergence between computer, telecommunications and cable television networks is generating turf wars between the Ministry of Information Industry (MII) and the State Administration of Radio, Film and Television (SARFT). At times, this conflict has become so heated that the personnel of these organisations have engaged in physical combat with each other. The stakes are high not only due to the economic interests being played for in what promises to be a lucrative market, but also due to the threat posed to the CCP's monopoly over the control of information presented by broadband convergence. While the political nature of this problem can be traced back to the weakness of the rule of law and the lack of clearly defined ownership in the PRC, technological development is what is creating the struggle in the first place.

Ngai-Ling Sum (Chapter 5) looks at the implications of the cross-border nature of Internet technology for the formation of politics and identity in the 'Greater China' region of southern Mainland China, Taiwan and Hong Kong. In doing so, she further deconstructs the state as the main agent of change by broadening the focus of analysis to include the formation of alliances between political leaders and corporate interests promoting a strategy of 'siliconisation'. The belief that ICTs are a panacea for the problems of economic development is, after all, a convenient theme to appeal to when promoting the development of a regional network of technological clusters modelled on California's Silicon Valley. From this perspective, political support has become increasingly important for entrepreneurs since fortunes have been lost in the wake of the Asian financial crisis and the bursting of the Internet bubble. The net result is a consolidation of the division of labour between different parts of the region and new tensions developing around old political fissures, such as Taiwan's problematic identity within China. Just as with Zhang, therefore, Sum's analysis indicates that the politics of siliconisation must be understood in terms of the complex relationship between technologically determined possibilities for new economic formations and the social responses that these generate from powerful actors.

Monika Ermert and Christopher R. Hughes (Chapter 6) extend the focus still wider, to the international arena, by looking at the way in which the technical nature of the global Domain Name System (DNS) generates political problems between states. Again, this case shows that it is technological developments that have forced states around the world to adapt to a kind of governance structure that many are unwilling to accept. It is the technological functions and specifications of the DNS that determine the way in which it has evolved into a highly centralised system in order to achieve the maximum scope for interoperability, and management of the system has to reflect this. Moreover, because the DNS has been physically developed by engineers based in the United States, its emergent system of governance has fallen under the jurisdiction of the American courts, and the Department of Commerce has an effective veto over any measures that might be taken to change the address system. With China having joined the Internet relatively late, it now faces the challenge of how to regain control over what the government sees as its rightful portion of cyberspace. Unless the technology itself can be decentralised, however, it is difficult to see how this can be achieved.

Christopher R. Hughes (Chapter 7) finishes the volume by looking at the complex ways in which the development of ICTs impacts on international security. China's policy-makers and leaders are certainly aware of the threat posed to the PRC's security by the necessity of relying on foreign, largely American, technology, as well as the ability of technologically advanced states to set the agenda when it comes to making international rules to govern new ICTs whenever they reach a new stage of development. In addressing these concerns, the thinking and actions of Chinese policy-makers tend to reflect the political tradition from which they emerge. This is most obvious in the promotion of an updated version of Mao's theory of the 'Three Worlds', according to which the information age has seen the appearance of information hegemony, information sovereign states, and information colonial and semi-colonial states. Moreover, just as in previous chapters, Hughes argues that the responses to such perceptions have to be understood as originating from a variety of powerful actors who can claim a stake in addressing the threat of information warfare, including the military, various ministries and industrial interests. This latter group, it is important to note, includes foreign as well as domestic firms. Moreover, the government certainly believes that it can work with the international community for the sake of mutual security, both by learning from foreign legislation about how to control and monitor activity on the Internet, and to work with other states who are equally threatened by the prospect of an unregulated global cyberspace. Of course, the likelihood that such cooperation will be forthcoming has been considerably raised by the terrorist attacks on New York and Washington that took place on 11 September 2001.

What all the contributors of this book have done, then, is to develop in their own ways methods for understanding how the political impact of ICTs is shaped by the society in which they are appropriated. The sociology of technology has long revealed that the impact of new technologies is determined as much by the ways in which they are appropriated and used by people, as by their intended design function. If this methodological standpoint is accepted, then understanding the impact of information technology in China requires knowledge not only of the ways in which technology works, but also of the political system, culture and history of that country.

It is not suggested anywhere in this work that China's 'digital leap forward' will have an outcome anything like as dramatic or tragic as Mao's experiments with collectivisation in the 1950s. The contents of this book do, however, serve as a reminder that if the political context is not right the resources being poured into ICTs by Chinese policy-makers may not be put to much more efficient use than the pots and pans that were melted down to make 'steel' in the backyard furnaces of the Great Leap Forward. So far, the ability of the state to appropriate advanced technology remains hampered by the political and social structures within which it is to be used. This includes the weakness of the rule of law, stark gaps of development between regions, obstacles posed by bureaucratic competition, the fear of losing control over domestic dissent and over the activities of foreigners. From a longer historical perspective such problems are symptomatic of the enduring crisis of identity that permeates the culture of a society still struggling to come to terms with the transition from a world civilisation to a nation state.

Notes

1 On the *'ti-yong'* formula, see J. Levenson, *Confucian China and Its Modern Fate*, Berkeley and Los Angeles: University of California, 1965, pp. 59–79.
2 N. Hachigian, 'China's Cyber-Strategy'. *Foreign Affairs*, vol. 80, no. 2, March/April 2001. Online. Available HTTP: <http://www.rand.org/nsrd/capp/cyberstrategy.html>; G. Walton, *China's Golden Shield: Corporations and the Development of Surveillance Technology in the People's Republic of China*, Montreal: International Centre for Human Rights and Democratic Development, 2001. Online. Available HTTP: <http://www.ichrdd.ca/frame.iphtml?langue=0> (accessed 29 October 2001); T. Boas and S. Kalathil, 'The Internet and State Control in Authoritarian Regimes: China, Cuba, and the Counterrevolution'. Carnegie Working Paper no. 21, July 2001. Online. Available HTTP: <http://www.ceip.org/files/pdf/21Kalathil Boas.pdf> (accessed 10 May 2002).
3 R.A. Caro, *The Power Broker: Robert Moses and the Fall of New York*, New York: Random House, 1974. See also L. Winner, 'Do Artifacts Have Politics?' in D. Mackenzie and J. Wajcman, *The Social Shaping of Technology* (Second Edition), Buckingham and Philadelphia: Open University Press, 1999, p. 30.
4 L. Winner, 'Do Artifacts Have Politics?', p. 37. For further examples of technologies shaping societies, see A.D. Chandler Jr, *The Visible Hand: The Managerial Revolution in American Business*, Cambridge, MA: Belknap, Harvard University Press, 1977.

1 ICTs in China's development strategy

Xiudian Dai

> [L]eapfrogging in productivity development may be achieved . . . by melding informatization and industrialization, the two processes reinforce each other and progress simultaneously.
>
> (PRC Premier Zhu Rongji, October 2000)[1]

At first sight, the belief of China's reformist Premier Zhu Rongji that informatisation is the way to achieve a developmental 'leapfrog' seems to resound with echoes of the days when China tried to catch up with the advanced industrial states under the leadership of Mao Zedong. However, much has changed since the days of centralised planning, the promotion of heavy industry and the collectivisation of farming within a virtually self-closed agricultural society, let alone mass political movements such as the Great Leap Forward and the Cultural Revolution. Since the end of the Mao era, development has been pursued through the more modest means of introducing market mechanisms into the planned economy to achieve comparatively modest aims, such as the Four Modernisations in agriculture, industry, defence and science and technology. It is within this reformist policy framework that ICTs have become increasingly salient as a strategic priority while the global information revolution has taken place.

Any belief in the revolutionary potential of ICTs amongst the Chinese leadership is thus as likely to have been influenced by and derived from the up-beat assessments that have been produced outside China as by the heritage of the country's own past. These range from Alvin Toffler's theory of the 'Third Wave'[2] to the host of academics and international organisations that have urged developing countries to take advantage of the global communications revolution to jump-start economic development.[3] It is the European Commission, for example, that makes the claim that 'ICT and new networks offer real opportunities and advantages to those who have the will and make the effort to use them, and these potentials should be actively pursued and not denied'.[4] Others warn that '[t]he economic future will belong to the technologically adept'.[5]

While many observers and policy-makers seem to think that the world's problems can be solved by simply spreading the 'digital opportunity' to get hooked up to networks, however, China's leaders might do well to heed those with a more conventional approach to development who favour concentrating on more traditional tools and methods.[6] Such cautious voices argue that the 'new economy' is a phenomenon confined largely to the United States, with limited relevance for other countries.[7] Developing economies, in particular, may be unable to catch up with the industrialised world in the 'information age' because 'technology is a reward of development, making it inevitable that the digital divide follows the income divide'.[8] If such pessimism is warranted, then 'leapfrogging' to a higher level of economic and social development on the back of ICTs will be impossible.[9] The debate on the relationship between ICTs and development is thus far from conclusive.

In some respects, the Chinese strategy for meeting the challenges of the information revolution can be seen as lying between these poles of utopianism and cynicism. As Zhu Rongji's statement makes clear, at its core lies the attempt to achieve an optimal combination of the 'new economy' and the 'old economy', or digitalisation hand-in-hand with industrialisation. This chapter will assess the feasibility of such an approach by looking at the ways in which government strategy and public policy have impacted on the process of infrastructure development, technological innovation, the take-up of new ICTs (in particular the Internet) and administrative efficiency in China.

Chinese optimism in the global context

Government policies and programmes to promote the development and application of ICTs in the context of general scientific and technological modernisation began to be initiated in China in the mid-1980s. The most significant of these was the '863' Programme (so-called because it was launched in March 1986),[10] which aims at promoting excellence in scientific research and the building of a national capacity in high technologies that can compete with the western industrialised countries. Under its umbrella, state research funds have been allocated to leading universities and institutions such as the Chinese Academy of Sciences, which are engaged in strategic research and development activities. Its projects embrace telecommunications, optical-electronics, artificial intelligence and information processing, as well as space technologies and biotechnologies. In many respects, the '863' Programme can be seen as the Chinese version of the SDI (Strategic Defence Initiative), Eureka and the Fifth Generation Computer in the industrialised countries.

The twin-track strategy of developing informatisation in parallel with industrialisation began somewhat later, during the period of political transition from the leadership of Deng Xiaoping to that of Jiang Zemin as the 'core' of the third generation of CCP leadership, which began after the government's crackdown of the Tiananmen protesters in June 1989 and ended with the death of Deng in 1997. It was at this time that the government launched its Informatisation of the National Economy (INE) programme.

More recently, the strategic thinking behind this project has been incorporated into the Tenth Five-Year Plan (2001–5), which makes informatisation of the national economy and society a strategic priority.[11]

The INE initiative was launched just as a consensus began to emerge among international organisations that the global information revolution could provide an opportunity for many, if not all, developing countries to leapfrog in both technological and economic development.[12] It was during the early 1990s that the European Commission actively began to promote ICTs as a strategic tool for economic growth in developing countries, seeing the information society as offering more efficient management for small and medium-sized enterprises (SMEs), enhanced provision of economic information, better training, interactive user/server networks, access to international markets and generally improved efficiency for government and administration.[13] The positive atmosphere surrounding the future of ICTs was encapsulated when leaders of the G7 countries, the European Commission and Southern Africa heralded the emergence of a global 'information society' that would have a profound impact on socio-economic development, when they met in Brussels in February 1995.

It is worth noting here that this international optimism has continued well into the new millennium, as was made clear when leaders of the western industrialised countries confirmed their belief that new ICTs present 'one of the most potent forces in shaping the twenty-first century' that are 'fast becoming a vital engine of growth for the world economy' when they met for the G8 summit (the G7 countries plus Russia) in Okinawa in July 2000.[14] Despite a note of caution over the need to address the existence of a 'digital divide' within and between countries, there was much to encourage Chinese leaders when the meeting made a declaration which concluded that countries that succeed in harnessing the potential of ICTs 'can look forward to leapfrogging conventional obstacles of infrastructural development, to meeting more effectively their vital development goals . . . and to benefiting from the rapid growth of global e-commerce' (G8, 2000). Even as late as February 2002, well after the bursting of the e-commerce bubble, the UNDP felt confident enough to sponsor a Global Digital Opportunity Initiative. When this was launched at the World Economic Forum in New York, the president of the Markle Foundation remained adamant that the appropriate deployment of new technologies can 'offer an unprecedented opportunity to meet global development challenges'.[15]

The technological factors feeding such optimism are not hard to find. The European Commission, for example, argues that moving towards the information society 'entails a reduction of time and space constraints and presents a panoply of new tools with unparalleled capacities enabling the developing countries to make some great leaps forward in technology by economizing on the intermediary stages which the industrialized countries have gone through'.[16] Table 1.1 illustrates the scale of the impact that innovation in ICTs has had on the costs of transmitting data. According to these figures, it

Table 1.1 The costs of communications

Model I: Vertical (historical) comparison	The cost of transmitting a trillion bits of information from Boston to Los Angeles has fallen from USD 150,000 in 1970 to 12 cents today.
Model II: Horizontal (technological) comparison	E-mailing a 40-page document from Chile to Kenya costs less than 10 cents, faxing it about USD 10, and sending it by courier USD 50.

Source: Data from UNDP, *Human Development Report 2001*.

is 500 times more expensive to deliver a 40-page document using conventional posts (courier) than it is by using e-mail.[17] In light of such statistics, it seems reasonable enough to believe that economically less-developed countries should be able to reap significant economic gains from innovations in the communications sector.

It has been in the context of this growing international consensus over the opportunities and challenges of the global information revolution, then, that the Chinese leadership has given increasing 'prominence to the development of science, technology and education', and acceleratation of 'the informatization of national economic and social progress'[18] to meet the priority of economic restructuring, to borrow the words of President Jiang Zemin. With the economic benefits of informatization depending on the existence of a competitive ICT sector, it is clear why this area of the economy has been made a priority in the current Five-Year Plan.

The priority given to informatisation by the government is made quite clear by the thinking of the Ministry of Information Industry (MII), which sees ICTs as a key sector of industry that has risen from making a total value-added contribution to GDP of 1.98 per cent during the Eighth Five-Year Plan (1991–5), to account for 3.4 per cent during the Ninth Five-Year Plan (1996–2000), and with expectations for it to notch up 6.7 per cent during the Tenth Five-Year Plan. While such figures may not be overwhelming, the MII explains that ICTs also play a strategic role because they provide the infrastructure necessary to support the growth of other industrial sectors in the information age. It thus calls ICTs a 'dragon head' (*longtou*), creating new economic dynamics that can help to promote the reform of more conventional industries and raise their productivity by reducing transaction costs. They are even seen as helping with efforts to protect the environment and achieve sustainable development by reducing material consumption.[19]

Head of the MII, Wu Jichuan, explains that the government's national strategy for the promotion of the ICT sector in the first ten years of the twenty-first century includes the following principal elements:

- Speeding up the construction of a new generation of ICT infrastructure alongside the strategic and structural adaptation of the current

infrastructure, to reflect the need for convergence between telecommunications networks, TV transmission networks (in particular cable TV) and computer networks (such as the Internet).
- Achieving breakthroughs in the core areas of ICTs in order to foster a competitive manufacturing sector with an indigenous supply of key components based on Chinese Intellectual Property Rights (IPRs). Breakthroughs are expected in areas such as large-scale integrated circuits (LSIC), high-speed and high-capacity computers, large-scale operating systems, super high-speed network systems, new generations of mobile communications and digital TV.
- The systematic promotion of the application of ICTs in every domain of economic and societal development, centred on the re-engineering of conventional industrial sectors through key initiatives such as 'Government Online', 'Enterprise Online' and 'Family Online'.
- Encouraging more foreign and domestic investment by improving the effectiveness of governance and implementation of the rule of law in the ICT sector.[20]

It is hard to deny that such ambitious objectives might be characteristic of Chinese government propaganda over the decades. However, it will be argued below that a proper appraisal must also acknowledge that the visions and optimism of the leadership are being translated into policies which have become a real driving force behind the quest for what might be called a 'new economy with Chinese characteristics'.

Broadband China: creating the infrastructure

The development of an information infrastructure is broadly recognised as having a profound and wide-ranging impact on economies and societies.[21] China has long seen efficient transportation as a key factor in determining the country's success in meeting international economic competition, and in the age of the Internet it has become just as crucial to be able to maximise the ability to move digital information in high volumes and to increase the value of that information on a national communications infrastructure. This has become even more true with the advent of broadband networks, the next generation of Internet service that offers to fuel a new wave of innovation. Fast and always on, broadband promises to deliver music and video content in new ways, and to lead to the development of applications that have not yet been imagined by offering innovators and creators a whole new platform on which to build.[22]

China has had to rise to this kind of technological challenge from a very low base. In 1980 there were only 8,000 km of long-distance cables catering for 22,000 long-distance telephone lines.[23] These were used almost exclusively by CCP organs, government agencies and the People's Liberation Army (PLA), making telephone services a luxury beyond the reach of ordinary citizens.

ICTs in China's development strategy 13

During the 1980s this inadequate provision of communications infrastructure became a big concern for the Chinese leadership. The MII claims that in March 1980 and February 1984 Deng Xiaoping himself gave talks on national economic development strategy in which he stressed the strategic importance of transportation and telecommunications for overall economic development.[24] Needless to say, from an institutional perspective, such a position fits snugly with the MII's agenda to oversee the rapid development of a communications infrastructure over which it will itself be the powerful guardian.

What is particularly interesting, however, is that this low level of telecommunications infrastructure provision can be seen as something of a latecomer's advantage, rather than as an obstacle to China's participation in the global communications revolution. This is because industrialised countries already possessing a sophisticated analogue copper wire network have to engage in a costly process of 're-engineering' with digital techniques (such as asynchronous digital subscriber line or ADSL technology) to achieve broadband capabilities. China, on the other hand, can make a 'jump-start' by deploying a nationwide fibre-optic cable network that offers a high capacity of digital transmission from scratch. The 1990s has indeed witnessed spectacular progress in the construction of information superhighways in China, with a total of 22 long-distance fibre-optic cables deployed by the mid-1990s (Table 1.2). By the end of 1998, a high-speed national grid of information superhighways was already in place, linking together all provinces and major cities. This national grid consists of eight 'horizontals' (long-distance fibre links cutting through the country east-westward) and eight 'verticals' (long-distance fibre links cutting through the country north-southward). The total length of fibre-optic cables in China had already reached 1.5 million km by the end of 2001,[25] representing a huge increase from 1980.

This completion of the national grid of fibre-optic cables in China within a relatively short space of time represents an impressive achievement by any standard. Moreover, moving straight into fibre-optic cables has been an obvious part of the 'leapfrogging' strategy that is being adopted by a growing number of developing countries that can largely skip the intermediate stages of copper wire and analogue telephone systems.[26]

In addition to this technological 'leapfrogging', the government has also stimulated the spread of broadband by breaking up the monopoly on telecommunications services that was held by China Telecom (owned by the former Ministry of Posts and Telecommunications) until a programme of telecommunications reform began in 1994. Following a recent decision by the government to split China Telecom into two independent companies, the telecommunications sector now has six major operators:[27]

- China Telecom (CT, based on the southern parts of the former China Telecom)

Table 1.2 China's information superhighways: fibre-optic cable links (1991–95)

Fibre-optic cable links	Traversing provinces	Length (km)	Date complete
Beijing–Chengde–Tongliao–Baicheng–Qiqihar	Hebei, Inner Mongolia Jilin, Heilongjiang	2,600	1995
Beijing–Hohhot–Yingchuan–Lanzhou	Hebei, Inner Mongolia Ningxia, Gansu	1,990	1995
Beijing–Shenyang–Harbin	Hebei, Liaoning, Heilongjiang	2,100	1995
Beijing–Tianjin–Jinan–Nanjing	Hebei, Shangdong, Jiangsu	950	1993
Beijing–Tianjin–Tangshan	Hebei	245	1994
Beijing–Wuhan–Guangzhou	Hebei, Henan, Hubei, Hunan, Guangdong	2,945	1994
Beijing–Taiyuan–Xian	Hebei, Shanxi, Shaaxi	1,720	1995
Chengdu–Chongqing–Guiyang–Changsha–Nanchang–Hangzhou	Sichuan, Guizhou, Hunan, Jiangxi, Zhejiang	4,500	1995
Chengdu–Guiyang–Hangzhou–Fuzhou	Fujian, Zhejiang, Jiangxi, Hunan, Guizhou, Sichuan	4,354	1995
Chengdu–Kunming	Sichuan, Yunnan	1,200	1993
Chengdu–Xian–Zhengzhou–Xuzhou	Sichuan, Shaaxi, Henan, Anhui, Jiangsu	1,700	1995
Chongqing–Wuhan	Sichuan, Hubei	790	1994
Fuzhou–Shanghai	Fujiang, Zhejiang	1,160	1993
Fuzhou–Guangzhou	Fujiang, Guangdong	1,336	1993
Guangzhou–Haikou	Guangdong, Hainan	858	1993
Guangzhou–Nanning	Guangdong, Guangxi	804	1993
Harbin–Changchun–Shenyang	Heilongjiang, Jilin, Liaoning	550	1995
Kunming–Nanning	Yunnan, Guangxi	1,200	1995
Nanjing–Wuhan	Jiangsu, Anhui, Hubei	980	1993
Nanjing–Shanghai	Jiangsu	280	1993
Shanghai–Wuhu	Jiangsu, Anhui	300	1994
Xian–Lanzhou–Urumqi–Xining	Shaaxi, Ningxia, Gansu, Qinghai, Xinjiang	3,150	1995

Source: Xiudian Dai, *The Digital Revolution and Governance*, p. 105, adapted from Ante Xu and P. Armstrong, *Chinese Telecom Market*, 1995, pp. 95–7.

- China Netcom Jitong (CNJ, incorporating the northern parts of the former China Telecom, the former China Netcom and the former Jitong Communications)
- China Unicom (CU, the first competitor to the former China Telecom)
- China Railcom (CR, owned by the Ministry of Railways)
- China Mobile Communications (CMC)
- China Satellite Communications Group (CSC)

The four fixed line operators – CT, CNJ, CU and CR – are the new proprietors of China's fibre-optic cable information superhighways. With this advanced information infrastructure in their hands, Chinese telecommunications operators are now in a position to provide not only voice telephony but also a variety of broadband services. Thanks to the completion of the national grid, Internet backbone operators are also able to increase the transmission capacity of their domestic and international backbone networks. For instance, ChinaNet (owned by China Telecom) has recently decided to launch a long-haul 10Gbps (Gigabits per second) system to provide a high-speed link between Shanghai and Hangzhou to meet the increasing demand for bandwidth.[28] This high-speed network will enable ChinaNet to deliver new services such as Internet protocol virtual private networks (IP-VPNs), voice over IP (VoIP) and video on demand (VoD), within and between the two major cities.

It is also important to note that the construction of the high-capacity national grid of fibre-optic cables has prompted a fast increase over the past few years in the bandwidth available for international Internet connection between China and the rest of the world (Figure 1.1). The total bandwidth of 7,597.5 Mb for international Internet traffic is currently shared by eight backbone network operators (Table 1.3), namely CSTNET (55 Mb, owned by the Chinese Academy of Science), CHINANET (6,032 Mb, owned by China Telecom), CERNET (257.5 Mb, owned by the State Educational Commission), CHINAGBN (168 Mb, owned by Jitong Communications), UNINET (418 Mb, owned by China Unicom), CNCNET (465 Mb, owned by China Netcom), CIETNET (2 Mb, owned by China International Electronic Trade Centre) and CMNET (200 Mb, owned by China Mobile). Furthermore, two additional backbone operators have also recently been approved by the government, namely CGWNET (owned by the China Great Wall Group) and CSNET (owned by ChinaSat).

Despite this rapid increase in bandwidth, however, it is also important to note that Internet traffic to and from sites outside China has to be channelled exclusively through international gateways located in the three cities of Beijing, Shanghai and Guangzhou. Although this arrangement might make government monitoring and control over information on the Internet easier, it also appears to be a major factor contributing to the slowness of Internet traffic in China. Until more international gateways are allowed, the real potential of increased bandwidth of the Internet backbone networks will not be fully utilised.

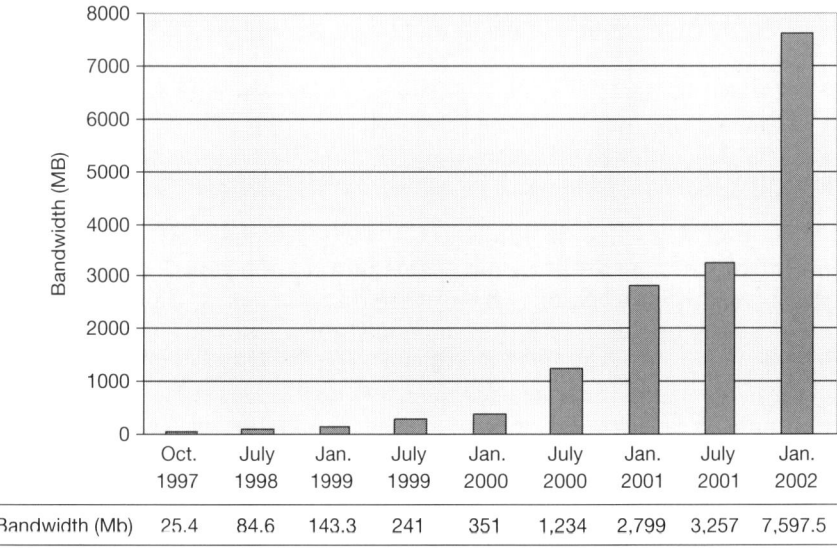

	Oct. 1997	July 1998	Jan. 1999	July 1999	Jan. 2000	July 2000	Jan. 2001	July 2001	Jan. 2002
Bandwidth (Mb)	25.4	84.6	143.3	241	351	1,234	2,799	3,257	7,597.5

Figure 1.1 Growth of total Internet bandwidth for international traffic (1997–2002)
Data: CNNIC, *Zhongguo hulianwangluo fazhan zhuangkuang tongji baogao (2002/1)*

Table 1.3 Internet backbone networks and their international bandwidth (January 2002)

Backbone network	Bandwidth	Owner	URL
CSTNET	55 Mb	Chinese Academy of Science	www.cstnet.net.cn
CHINANET	6,032 Mb	China Telecom	www.chinatelecom.com.cn
CERNET	257.5 Mb	State Educational Commission	www.edu.cn
CHINAGBN	168 Mb	Jitong Communications	www.gb.com.cn
UNINET	418 Mb	China Unicom	www.chinaunicom.com.cn
CNCNET	465 Mb	China Netcom	www.cnc.net.cn
CIETNET	2 Mb	China International E-Trade Centre	www.ciet.net
CMNET	200 Mb	China Mobile	www.chinamobile.com
CGWNET	–	China Great Wall Group	www.cgw.net.cn
CSNET	–	ChinaSat	www.chinasat.com.cn

Source: Based on CNNIC, *Zhongguo hulianwangluo fazhan zhuangkuang tongji baogao (2002/1)*; CNNIC, *Zhongguo hulianwangluo daikuan baogao* (*China Internet Bandwidth Report*).

Another benefit accruing from the new communications infrastructure centred on the national grid of fibre-optic cables is that China now has the world's second largest telephone network, boasting 179 million fixed lines.

This volume is second only to the United States, represents a teledensity of 13.9 per cent, and compares with a mere 10 million lines in 1992. In addition to this, China has already superseded the United States to become the largest mobile phone market in terms of its subscription base, with 145 million users in January 2002, or an 11.3 per cent penetration rate nationwide. The government's goal, as projected in the Tenth Five-Year Plan, is to increase the total number of telephone users to 500 million (comprising fixed line users and mobile phone users) by 2005. If this is achieved it will provide 15 per cent of the population with connections to the Internet (200 million users).[29] The ultimate aim is to build on the capacity of the high-speed broadband communications infrastructure to make China the largest Internet country in the world.[30] Some estimate that China will overtake countries such as Germany, Britain and South Korea to become the world's third largest broadband market as soon as the year 2006, trailing only the United States and Japan.[31]

A dramatic increase in the levels of investment in the telecommunications sector lies behind these infrastructure developments, amounting to a total of RMB 600 billion (bn) (USD 72.3 bn) since the founding of the PRC in 1949. The majority of this, however, has been during the past two decades. In fact, the period 1949–83 accounts for only RMB 6 bn.[32] In other words, China has been investing on average around RMB 31 bn (USD 3.7 bn) per annum since 1994 in its telecommunications network, which is a substantial undertaking for a developing country. Moreover, such investment has been further enhanced by the way in which telephone users have had to heavily 'subsidise' operators, mainly China Telecom, through measures such as the imposition of an 'initial installation fee (IIF)'. In the early 1990s, the government guideline for the IIF was RMB 3,000–5,000 (USD 600–1,000).[33] Compared to the present per capita GDP of USD 800, this was excessively high. Although the IIF was scrapped on 1 May 2001, this was a heavy cost to impose on many of the early 'movers' in the sector. The same can be said of high call charges. In early 1995, a 6-minute call from Shandong Province to the UK cost RMB 240 (or RMB 40 per minute). This is equivalent to USD 4.8 per minute at today's exchange rate.[34] In short, the rapid growth of the Chinese communications infrastructure is to a large extent the outcome of heavy government investment and user 'subsidies'. Of course, it could be added that such subsidies might well be detrimental to other types of spending.

'Hi-tech' China: searching for indigenous technologies

An advanced national grid of information superhighways would be of no use at all in the absence of 'goods' and 'vehicles' to run on it. China's new communications infrastructure is thus complemented by a drive to target research and development on key technologies and products in the ICT sector, with the ultimate aim of increasing indigenous ownership of these technologies. This effort is manifested in China's active involvement in the process of global standardisation and product innovation, where public policy is helping to increase the technological and manufacturing capacity of the sector. This fuels the optimistic view that '[T]he market can't do anything but grow in the coming years, particularly given that the Chinese government has put so much emphasis on the communications industry as a significant factor for overall economic growth'.[35]

It should also be noted that an important motive for government sponsorship of the development of new ICTs is a preoccupation among state organs with the role of ICTs in the shaping of global power politics. In the words of the MII:

> Various information technologies and standards have already become an important indicator of each country's economic sovereignty in the world. The ICT sector has become a strategic industrial sector for all countries to compete for the dominance in science and technology, the economy and military affairs. Therefore, developing the ICT sector is of great significance.[36]

For Chinese policy-makers, the ever-extending horizon of globalisation driven by cross-border flows of information and communications makes a country's total reliance on foreign hardware and software a worrying situation, especially when such equipment is American. The fast increase in the level of intelligence and capacity offered by new microchip designs, for example, has contributed not only to the creation of ever more powerful computers but also to the reduced control by human beings over their own information. The outcome of major state-sponsored projects to put government and enterprise online in China, will only exacerbate this problem by encouraging more and more state agencies and companies to connect their databases to the Internet. In theory, this has opened the floodgates for information to flow across the country's international borders. Unwilling and unable to shut down all channels of information, the government has thus opted for a strategy that promotes the development of indigenous technology in the hope that future equipment and systems will be designed and manufactured with an increasing proportion of 'trusted' components supplied by home-grown firms.

Fulfilling this is also attractive due to the sentimental value, or patriotic pride, that is generated by each technological breakthrough. China has already claimed a number of such advances, in fact. In July 2001, for example, it was announced to the world that Chinese engineers had successfully developed their first Central Processing Unit (CPU), a 32-bit microchip christened the 'Fangzhou-1'. Although it is still premature to suggest that this will enable the nation to take on world giants like Intel, this kind of indigenous research and development not only promises a wide range of applications for the design of ICT equipment, but just as importantly constitutes a significant source of national pride.[37] Coinciding with the launch of the 'Fangzhou-1', the State Development Planning Commission (SDPC) and the MII jointly decided to establish ten national centres of software development throughout the country, in the cities of Beijing, Shanghai, Dalian, Chengdu, Xi'an, Jinan, Hangzhou, Guangzhou, Changsha and Nanjing. These new state-funded bases are expected to play a key role in jump-starting development in the Chinese software industry.[38] While a large number of software houses do exist in China already, most of these are small firms that are unable to compete against global leaders such as Microsoft. Among them, only six have a sales revenue of more than RMB 1 bn (USD 120 million).[39] It is to remedy this situation that the Chinese government has made supporting the development of indigenous computer software an issue of strategic importance.

The government's desire for China to become a key player in the digital age is also demonstrated by its response to the process of global standardisation for 'third generation' (3G) mobile communications. China has in fact already developed its own 3G standard, known as 'TD-SCDMA' (time division-synchronous code division multiple access),[40] funded by the government to the sum of RMB 200 million (USD 24 million), in partnership with Siemens, which contributed a further USD 40 million.[41] It was accepted by the International Telecommunications Union (ITU) in March 2001 as a bona-fide 3G standard technology. Alongside the European W-CDMA (wide-code division multiple access) and the American CDMA-2000, countries throughout the world now have a third choice when it comes to selecting a standard for 3G mobile communications.

For China, the TD-SCDMA standard has transformed the country's image of itself as being a mere follower of foreign technology to become a technology proprietor in digital mobile communications. Regardless of whether TD-SCDMA is actually adopted by anybody, such a breakthrough is significant in itself for its symbolic value, since the Chinese mobile communications market has been dominated by western technologies, with leading mobile phone suppliers such as Motorola, Nokia and Ericsson engaged in fierce competition for a share. Moreover, the Chinese mobile phone market had developed to meet the European GSM (global standard for mobile communications) standard, until the beginning of 2002 when China Unicom launched a competing network based on the American CDMA standard. If nothing else, this coexistence of two competing western standards attested to the size of the Chinese market! With the emergence of TD-SCDMA, though, China is now poised to end this dependence on western standards and technologies in mobile communications. TD-SCDMA thus sends out a powerful message to the international community about the ability of China to leapfrog forward in certain areas of technological development, with any future decision on this issue likely to have a major impact not only on the Chinese market, but also on the global market for mobile communications.

In short, there are multiple motives behind the Chinese search for indigenous technologies and standards, ranging from economic considerations to matters of strategic security and national pride. The Chinese push for the indigenous ownership of ICTs thus echoes the view of commentators who remark that, '[a]lthough foreign technologies and expertise under the right circumstances can contribute to improved social and economic conditions in Third World countries, abundant evidence suggests that no recipient country can rely on it, whether the state is socialist or capitalist'.[42] China is certainly one of the Third World countries that is not content with relying on foreign technologies for building a digital economy.

Implications of e-commerce

Alongside the rapid development of the telecommunications sector, China has also experienced an exponential growth in the Internet since the mid-1990s. Although the proportion of the Chinese population online is still very low compared to many industrialised countries, the absolute number of 'netizens' increased from 620,000 in October 1997 to 33.7 million in January 2002.[43] This growth has been heavily driven by commercial interests, with the latest survey data from the China Internet Network Information Centre (CNNIC) indicating that 77.8 per cent of Chinese Websites (215,779 out of a total of 277,100) are dot-com sites. The geographical location of Websites suggests that a high proportion are concentrated in economically prosperous areas. In fact, Beijing, Shanghai and Guangdong Province account for 48 per cent of all of China's Websites registered under the Top Level Domain name '.cn'. Similarly, the majority of Chinese netizens are from comparatively developed regions, with 60.3 per cent located in eastern and south-eastern China, compared to only 4.6 per cent in north-western and 9 per cent in south-western China.[44] Bearing in mind that the government has recently begun a campaign to speed up economic development in central and western China, it is hoped that this internal 'digital divide' will be narrowed down when more organisations and individuals can afford to go online in the less developed regions.[45]

Despite the above figures, however, it must be admitted that the Internet sector is still at a very early stage of development in China and its direct contribution to the national economy should not be exaggerated. Nevertheless, its importance for development may be magnified by the way in which the adoption and exploitation of ICTs by existing industries has a broader impact on the process of restructuring.[46] Such a view is in fact reflected in state policy, where the government declares that it 'will work hard on e-commerce, accelerate the process of information and support enterprises in applying modern information network technology to international cooperation and exchanges'.[47] E-commerce certainly remains attractive to the government as a way to make up for China's lack of an efficient market-oriented distribution system. If widely adopted, it could be used by businesses and consumers to cut through the fragmented and multi-layered distribution system that has been created by decades of centralised economic planning.

Major B2B (business-to-business) dot-com sites, such as Alibaba.com and MeetChina.com, have already been successful in attracting both corporate clients and venture capital investment and the volume of e-commerce has reached a significant level in China. It is reported that, in the year 2000, a total of USD 47.17 million was transacted in business-to-consumer (B2C) transactions, with USD 9.29 bn in business-to-business (B2B). Optimistic forecasts suggest that B2B trading volume will increase by 22.8 per cent to USD 11.39 bn, while B2C transactions will climb 233.3 per cent to USD 157.25 million in 2001.[48] Even in the countryside there are reports of farmers living in remote areas successfully exploiting the economic potential of new ICTs. One farmer in south-east China is reported to have reached a million-dollar deal with a European customer via e-mail three years ago, while another is said to have sold 1,500 peacocks over the Internet by posting online advertisements in 2001. The Internet is claimed to have helped farmers in Zhejiang Province sell their produce to over twenty countries and regions in the world. In the coastal city of Ningbo, Zhejiang Province, over 30,000 farmers are said to be selling their produce on the Internet.[49] There is, therefore, already some evidence from China's experience with e-commerce that informatisation can help transform the industrial sectors of the 'old economy', including agriculture, and bring significant benefits for a developing country.

Governance of the information age

As digitalisation has become increasingly central to development strategy, it has become important for the government to minimise the possibilities for macro-level confusion in policy making and regulation concerning the ICT sector. To this end, in 1994 the government set up the Joint Conference for the Informatization of the National Economy (JCINE), headed by former Vice-premier Zou Jiahua. At its launch, the JCINE consisted of representatives from a total of twenty-four government ministries and state commissions, a number that in itself indicates the complexity of the Chinese bureaucracy that is pertinent to the sector. In 1996, the JCINE was replaced by the Information Technology Development Leading Group (ITDLG), under the same leadership. The mission of the ITDLG, similar to that of its predecessor, is to 'deal with the battling ministries'.[50] At present, building on the legacy of the JCINE and the ITDLG, another supra-ministerial body is in the making as the government is intent on establishing a State Council Information Management Commission (SCIMC). It is believed that this cabinet-level organisation will be modelled on the United States Federal Communications Commission (FCC) and will probably be under the leadership of Premier Zhu Rongji and Vice-president Hu Jintao,[51] the likely successor to Jiang Zemin as general secretary of the CCP.

In parallel with the establishment of *ad hoc* commissions at the State Council level to coordinate policies and strategies, the government has also undertaken sweeping ministerial reforms, something that has become increasingly important as turf wars have broken out due to technological convergence. These reached a new height in March 1998, when the State Council decided to merge the previously warring Ministry of Posts and Telecommunications (MPT), the Ministry of Electronic Industry (MEI) and the communications networks of three organisations including the Ministry of Radio, Film and Television (MRFT), China Aerospace Industry Corp., and China Aviation Corp. The outcome of this mega 'institutional merger' was the creation of the MII, widely regarded as a bold measure to address the challenges of technological convergence. To be sure, institutional and regulatory reforms are also hotly debated issues in western industrialised countries. For instance, the British government is developing plans to establish a new Ofcom (Office for Communications) to incorporate the UK's current telecommunications watchdog Oftel (Office for Telecommunications) and the ITC (Independent Television Commission) in order to replace its currently sector-specific regulatory structure with one that is more competent in dealing with the challenges brought by the digital revolution and, in particular, digital convergence. However, while top-level (State Council) coordination and institutional reforms have helped to eliminate turf wars between rival ministries in China, it is hard to claim that an effective structure of governance has been developed to cope with the many challenges associated with the all-embracing process of informatisation.

The problems that are arising in this respect can be seen, for example, in the controversial relationship between the three types of networks that are emerging, namely telecommunications, TV broadcasting (especially cable television) and computer networks. Zhang Chunjiang, Deputy Minister of Information Industry, has argued that under the impact of digital convergence, the sectoral borders between these previously separated networks no longer apply and that there should be no discrimination between networked services carried by television, telecommunications services and computer networks in the digital era.[52] However, such a view is vehemently rejected by parties with a vested interest in these networks. Among others, officials from the State Administration of Radio, Film and Television (SARFT) are quick to reject Zhang's remarks by arguing that television broadcasting enjoys a natural monopoly throughout the world.[53] Any move towards the convergence of the three independent networks is thus seen by the SARFT as an erosion of the 'natural monopoly' that it has hitherto enjoyed. Likewise, the MII has erected 'policy hurdles' to make it difficult for any 'external forces', including the SARFT, to enter into the potentially lucrative business of telecommunications services.[54]

The key network operators and service providers have also shown a distinct lack of interest in developing a single and converged transmission infrastructure, no doubt due to the fact that each one of them is already a big player with its own large customer base. For instance, the telecommunications industry now caters for 179 million fixed line users plus 145 million mobile users (the second and first largest market respectively in the world). The cable TV industry has also seen rapid growth in its customer base, with

17 per cent (or 80 million) of Chinese households now connected to the cable TV network, making China the biggest single market in the world. Meantime, the exponential growth of the number of Internet users in China has made it the most eye-catching new sector of the economy. Technically speaking, because each of the three big networks has the capacity to provide multimedia communications including voice telephony, data services and digital video, there is every reason to believe that customers would benefit from an integrated digital communications infrastructure. In reality, however, there might be a long way to go before we witness convergence between the three kinds of network in China.

In addition to resistance from the big operators in each sectoral domain, however, the biggest barrier to network convergence is believed to be public policy itself. For example, current state regulations stipulate that TV broadcasting organisations are not allowed to operate telecommunications services. Likewise, telecommunications operators are not allowed to be involved in broadcasting TV programmes.[55] This lack of a coordinated approach towards the development of the three individual networks has resulted in the construction and coexistence of different broadband networks competing for the same group of broadband customers at the local community level. For instance, Great Wall Broadband has invested RMB 357.5 million (USD 43 million) in the city of Guangzhou for its own broadband infrastructure and had attracted 224,000 broadband Internet customers by May 2001. In the same city, though, Guangdong Cable TV has also laid down its own cable network offering broadband services to 272,000 households. Meantime, Guangzhou Telecom (the local branch of China Telecom) has deployed a city-wide high-speed fibre-optic grid with a total bandwidth of 150Gb to offer broadband services to every part of Guangzhou.[56] A similar pattern of broadband infrastructure development can be seen in many other parts of China, with hundreds of operators competing to lay down their own networks. This nationwide and uncoordinated rush for broadband deployment earned 2001 the title 'China's Broadband Year'![57]

Such a lack of coordination has also led to concerns among government officials over the lack of technical standardisation. Song Ling, Director of the Office of the State Council Informatization Leading Group, recently commented that China's rush to promote informatisation has resulted in different sectors, government agencies and enterprises deploying information systems that are often incompatible with each other, leading to an inability to share resources effectively that is known as 'informational compartmentalisation' (*xinxi geju*).[58] Dealing with such problems has not been made any easier by the fact that a long-awaited telecommunications law is still in the drafting process. This work was originally the task of the former MPT, but it is a widely held view that the absorption of that ministry by the MII in 1998 did little to speed up the process of telecommunications legislation.[59]

A trickier problem is that the rules that do exist do not necessarily work as expected. For example, before China joined the World Trade Organisation

(WTO), state regulations barred foreign carriers from entering the Chinese telecommunications market, prohibiting direct investment in ownership, operations and the management of telecommunications networks. However, it has been considered 'normal' practice for some time to allow 'anomalies' to occur. Foreign participation in telecommunications can thus be indirectly achieved through the special arrangement of joint ventures with local partners.[60] Similarly, domestic regulations governing satellite broadcasting stipulate that foreign TV services can only be received in foreign compounds (such as embassies) and international hotels with a three-star rating or above. Despite these restrictions, however, Phoenix Satellite Television, owned by Rupert Murdoch's Star TV network with headquarters in Hong Kong, claims that it beams its Mandarin Chinese channel into 45 million homes in the mainland. The only explanation for this 'anomaly' is that Murdoch appears to have worked out a gentleman's agreement with the Chinese authorities, who seem happy to turn a blind eye.[61]

While the staggered timetable of market liberalisation agreed for China's entry into the WTO has helped to clarify the situation of foreign involvement, it is still uncertain what the ultimate effects of this process will be in practice. Moreover, an equally serious problem is posed by excessive regulation. This is manifested particularly strongly in the area of Internet governance, where government regulations introduced at different levels of the state apparatus impose a mass of specific requirements that have to be met by Internet service providers (ISPs), users and – most restrictive of all – content providers. The large number of decrees and their detailed nature makes China one of the most heavily regulated countries in the world when it comes to ICTs.[62] Bearing in mind the widely held conviction that contents are king in the digital age, this heavy-handed approach of the Chinese government towards media and communications in general is in danger of throwing out the baby with the bathwater by attempting to confine its citizens to officially defined 'information frontiers'. After all, informatisation can hardly be achieved by having no access to information!

Conclusion

The Chinese government's economic development strategy since the early 1990s has been characterised by an attempt to marry the opportunities and dynamics of the global communications revolution with the country's unfinished process of industrialisation. It has been argued above that there is already evidence to suggest that the twin-track strategy of promoting informatisation of the national economy in parallel with industrialisation can encourage the 'new economy' and the 'old economy' to reinforce each other in beneficial ways, provided appropriate public policy is in place. After several years of heavy investment in new infrastructure and promoting the application and development of new ICTs, China has emerged as a substantial player in the digital wave. Among the country's noteworthy achievements are the building of an advanced communications infrastructure with the world's second largest fixed line telephone network, largest mobile phone network and largest cable TV network. Such developments indicate that the process of 'leapfrogging' by developing countries in certain areas of technological and economic development is not impossible. The unprecedented opportunities provided by new ICTs, in particular the Internet, for re-engineering the 'old economy' can thus be considered a serious option for policy-making in developing countries, although practice will always be affected by local historical, economic and social characteristics. Overall, though, the Chinese government can be said to have displayed a great deal of vision in its policy of closing the gap with the industrialised countries by placing digital communications technologies at the heart of its development policy.

It has also been argued, however, that there remain some decidedly problematic factors in the Chinese government's approach to informatisation, which impose heavy burdens on both the state and consumers when it comes to carrying through and financing the development of the National Information Infrastructure (NII). A particular drawback lies in the weak structures of governance over the ICT sector. To be sure, China is one of the most pioneering countries in the developing world when it comes to undertaking substantial reforms in the governance of telecommunications and other aspects of new ICTs. Since 1994, this has included deep structural reforms to the telecommunications sector and ministerial re-organisation to create the Ministry of Information Industry. Plans are in place to take this institutional reform further in the near future to meet the challenges posed by digital convergence. Yet, despite all this, the long-awaited telecommunications law has not yet been enacted and government regulations are often by-passed by convenient 'anomalies'. Now that China is member of the WTO, the government will be under increasing pressure to address these issues in line with international obligations. The continued success of China's informatisation programme would seem to be contingent upon the ability of the government to meet these obligations.

As a final note, it should be added that the Chinese government's active pursuit of a first-rate infrastructure and advanced technologies has not been matched by an equally proactive policy to promote the participation of citizens and free up digital content provision. Despite initiatives such as 'Government Online', 'Enterprise Online', 'Family Online' and the promotion of e-commerce, government policies and regulations tend to be restrictive rather than enabling in terms of content provision and access to information. Ultimately, no matter what the technology is, it is evident that political considerations tend to outweigh economic benefits when it comes to policy-making for the new economy, in just the same way that they have always triumphed over the old economy. This will undoubtedly be detrimental to the long-term economic exploitation of the potential that the new information infrastructure and new technologies can offer China.

Notes

1 Speech by Zhu Rongji, '*Guanyu zhiding guomin jingji he shehui fazhan di wuge wunian jihua jianyi de shuoming*' ('Remarks about the Drafting of the Tenth Five-year Plan for the Development of the National Economy and Society'), 9 October 2000, in *People's Daily*, 19 October 2000. Online. Available HTTP: <http://www.peopledaily.com.cn/GB/channel1/10/20001019/27862.html> (accessed 20 October 2000).
2 A. Toffler, *The Third Wave*, London: Pan Books, 1981.
3 The idea of jump-starting economic development in the information age is advocated by a growing body of literature including, among others, the following: H. Dordick and Georgette Wang, *The Information Society: A Retrospective View*, London: Sage, 1993; B.R. Scott, 'The Great Divide in the Global Village', *Foreign Affairs*, January/February 2001, pp. 160–77; European Commission, *The Information Society and Development: A Review of the EC's Experience in Asia, Latin America and the Mediterranean*, DG External Relations, ER/04 Economic Analysis, Brussels, 12 January 2001; European Commission, *The Information Society and Development: The Role of the European Union*, Communication to the Council and the European Parliament, COM(97) 351 final, 15 July 1997. Online. Available HTTP: <http://europa.eu.int/ISPO/intcoop/i_com_97_351.html> (accessed 15 February 2002); World Bank, *China's Development Strategy: The Knowledge and Innovation Perspective*, Washington DC: World Bank, 2000; World Bank, *Knowledge for Development*, Washington DC: World Bank, 1998/99.
4 European Commission, *The Information Society and Development: A Review of the EC's Experience in Asia, Latin America and the Mediterranean*, 2001, p. 37.
5 Angus King, Governor of Maine state, quoted in K. Dean, 'Maine Students Hit the Ibooks', *Wired News*, 9 January 2002. Online. Available HTTP: <http://www.wired.com/news/school/0,1383,49046,00.html> (accessed 10 January 2002).
6 European Commission, *The Information Society and Development*, p. 37.
7 OECD, *Is There a New Economy? First Report on the OECD Growth Project*, Paris: OECD, June 2000; 'Waiting for the New Economy', *The Economist*, 14 October 2000, pp. 70–1.
8 United Nations Development Programme (UNDP), *Human Development Report 2001: Making New Technologies Work for Human Development*, Oxford: Oxford University Press, 2001, p. 27.
9 A. Persaud, 'The Knowledge Gap', *Foreign Affairs*, March/April 2001, p. 108.
10 The date 'March 1986' is often written as '86.3' in Chinese.
11 State Council, *Shiwu jihua gangyao quanwen* (*Complete Text of the Outline of the Tenth Five Year Plan*). Online. Available HTTP: <http://www.chinaemb.or.kr/chn/9272.html> (accessed 10 April 2002).
12 For more discussions about the view that developing countries should seek to leapfrog in technological and economic development, see UNDP, *Human Development Report 2001*; World Bank, *China's Development Strategy*; World Bank, *Knowledge for Development*.
13 European Commission, *The Information Society and Development: The Role of the European Union*, 1997.
14 G8, 'Okinawa Charter on Global Information Society', Okinawa, 22 July 2000. Online. Available HTTP: <http://www.library.utoronto.ca/g7/summit/2000okinawa/gis.htm> (accessed 13 October 2000).
15 J. Krane, 'U.N. Looks to Narrow Tech Gap', *Yahoo! News*, 5 February 2002. Online. Available HTTP: <http://dailynews.yahoo.com/h/ap/20020205/tc/global_tech_gap_1.html> (accessed 5 February 2002).

16 European Commission, *The Information Society and Development: The Role of the European Union*.
17 The mathematics is rather simple: the accumulated cost of 10 deliveries of a 40-page document by courier from Chile to Kenya could well be enough to buy a computer!
18 Jiang Zemin, untitled address at the dinner for the opening of the Fortune Global Forum: Next Generation Asia, 8 May 2001, Hong Kong. Online. Available HTTP: <http://www.timeinc.net/fortune/conferences/Global/pjztrans.html> (accessed 23 January 2002).
19 Official policy statement by the Ministry of Information Industry, '*Erlun xinxi chanye bumen ruhe guanche wuzhong quanhui jingshen*' ('On How to Implement the Spirit of the CCP Fifth Plenary Meeting within the Information Industry Agencies (2)'). Online. Available HTTP: <http://www.ceiinet.gov.cn/00/news/200011/09/093920.asp> (accessed 13 November 2000).
20 See Yi Zhang, '*Wu Jichuan toulou Zhongguo xinxi chanye fazhan zhanlüe*' ('Wu Jichuan Reveals Information Industry Development Strategy'), *People's Daily*, 9 February 2001.
21 OECD, *Special Issue on Information Infrastructure*, STI Review, no. 20, Paris: OECD, 1997, p. 7.
22 L. Lessig, 'Who's Holding Back Broadband?', *Washington Post*, 9 January 2002. Online. Available HTTP: <http://www.newsbytes.com/news/02/173496.html> (accessed 10 January 2002).
23 Xiudian Dai, *The Digital Revolution and Governance*, Aldershot: Ashgate, 2000.
24 Ministry of Information Industry, '*Woguo dianxin lingyu gaige kaifang dashiji*' ('Milestones of Reform and Opening up in Our Country's Telecommunications Sector'). Online. Available HTTP: <http://www.mii.gov.cn/> (accessed 8 February 2002).
25 Wu Jichuan, Information Industry Minister, '*Yingjie quanqiu xinxi wangluohua de xin tiaozhan*' ('Embracing the New Challenges of the Global Networked Information'), 14 January 2002. Online. Available HTTP: <http://www.mii.gov.cn> (accessed 8 February 2002).
26 World Bank, *Knowledge for Development*, p. 9.
27 The decision to settle for six operators was announced by the government on 11 December 2001, the same day that the PRC won accession to the WTO. The date for the launch of the new China Netcom operator was set as 17 May 2002.
28 The rapid increase in demand for bandwidth in the Shanghai–Hangzhou area is associated with the fact that Hangzhou has now become the most favoured residential city for many of those who work in the Pudong area of Shanghai and command a high salary.
29 Wu Jichuan, '*Yingjie quanqiu xinxi wangluohua de xin tiaozhan*'.
30 According to one survey, by March 2002 China had already become the world's second-largest Internet nation, with some 56.6 million people having home Internet access in China, second only to the US. For more details see D. Kelsey, 'China At-Home Net Head Count No. 2 in World', *Newsbytes*, 22 April 2002, <http://www.newsbytes.com/news/02/176049.html> (accessed 23 April 2002).
31 M. Donegan, 'Eastern Promise', *Communications Week International*, Issue 278, 4 February 2002, pp. 12–14.
32 See Dai, *The Digital Revolution and Governance*, p. 100; Feng Kuang, '*Dianxinwang jisuanjiwang guangbodianshiwang sanwang ronghe: xiayidai hulianwang zhongjie dianxin "Sanguo shidai"*' ('The Convergence of Telecoms Networks, Computer Networks and TV Broadcasting Networks: Next Generation Internet to End the "Era of Three Kingdoms"'), *Beijing Evening News*, 16 August 2001.

28 *Xiudian Dai*

33 Note that the official exchange rate was about USD 1 for RMB 5 in the early 1990s.
34 Today's exchange rate is USD 1 for RMB 8.3. This rate is used throughout the chapter, unless indicated otherwise. Information on the call charges from Shandong to the UK is from the author's personal communication in September 2000 with a local resident, who actually made the call and paid the bill.
35 M. Donegan, 'Eastern Promise', p.12.
36 Ministry of Information Industry, '*Erlun xinxi chanye bumen ruhe guanche wuzhong quanhui jingshen*'.
37 At the time of writing there is no report about any actual production of the Fangzhou-1 chips in China.
38 Xinhua News Agency, '*Wu Bangguo zhichu: Nuli tuidong woguo ruanjian chanye shixian kuayueshi fazhan*' ('Remarks by Wu Bangguo: Endeavouring to Achieve Leapfrogging in the Development of the Software Sector in Our Country'), *People's Daily*, 12 July 2001.
39 Jianrong Di, '*Woguo queding 10 ge guojiaji ruanjian jidi: Wu Bangguo yaoqiu ruanjian chanye shixian kuayueshi fazhan*' ('Our Country to Have 10 National Software Centres: Wu Bangguo Urges the Software Sector to Achieve Leapfrogging in Development'), *Jiefangjun bao (PLA Daily)*, 13 July 2001, p. 2.
40 TD-SCDMA has been jointly developed through a partnership led by the Chinese Academy of Telecommunications Technology, Datang (a Chinese firm) and Siemens of Germany.
41 Investment figures are from *Chinabyte*, 13 July 2001, cited in '*Zhongguo disandai yidong tongxin keneng youxian caiyong TD-SCDMA jishu*' ('China May Give Priority to TD-SCDMA Technology for Its 3G Mobile Communication'), *People's Daily*, 13 July 2001.
42 G. Sussman, *Communication, Technology and Politics in the Information Age*, London: Sage, 1997, p. 248.
43 CNNIC, *Zhongguo hulianwangluo fazhan zhuangkuang tongji baogao (2002/1)* (*China Internet Development Statistics Report (2002/1)*). Online. Available HTTP: <http://www.cnnic.net.cn/develst/2002-1/> (accessed 18 January 2002).
44 Figures are from CNNIC, *Zhongguo Hulianwangluo Fazhan Zhuangkuang Tongji Baogao*.
45 For a more detailed account on the 'digital divide' in China see Chapter 2 by Karsten Giese in this volume.
46 OECD, *Is There a New Economy?*, p. 9.
47 Jiang Zemin, untitled address at the dinner for the opening of the Fortune Global Forum.
48 Figures are cited by the CNN.COM, 'China Farmers Peddle Produce Online'. Online. Available HTTP: <http://www.cnn.com/2001/BUSINESS/asia/07/16/hk.wiredfarmers/index.html> (accessed 23 January 2002).
49 Ibid.
50 Tania Chau, J. Goldstein, A. Hamilton and M. Scram, 'Cyber Elite – Zou Jiahua', *Time*, 22 September 1997, pp. TD14–TD15.
51 See J. Kynge, 'China Looks to Ring Changes in Telecoms Sector', *Financial Times*, 17 October 2001, p. 17.
52 See remarks by Zhang Chunjiang on the convergence of telecommunications networks, computer networks and TV broadcasting networks, cited in Feng Kuang, '*Dianxinwang jisuanjiwang guangbodianshiwang sanwang ronghe: xiayidai hulianwang zhongjie dianxin "Sanguo shidai"*'.
53 See *Zhongguo xinwen she* (China News Agency), '*Xinxi chanye bu yankong hezizhe "ruqin" dianxin shichang*' ('MII's Tight Control over the "Invasion" in the Telecommunications Market by Joint Investors'), *People's Daily*, 23 August 2001.

54 Ibid. For a more detailed discussion on the 'turf war' between the MII and the SARFT, see Chapter 4 by Junhua Zhang in this volume.
55 Feng Kuang, *'Dianxinwang jisuanjiwang guangbodianshiwang sanwang ronghe: xiayidai hulianwang zhongjie dianxin "Sanguo shidai"'*. Note that in certain western industrialised countries such as Britain, established state regulations also bar telecommunications operators from entering TV broadcasting services.
56 *'Guangzhou kuandai jieru 55 wan hu'* ('Broadband Connections Reached 550000 Households in Guangzhou'), *People's Daily*, 4 June 2001. Online. Available HTTP: <http://www.peopledaily.com.cn/GB/it/49/150/20010604/481273.html> (accessed 18 December 2001).
57 Cited in Ke Liu, *'Zhongguo kuandai: shenghuo bu "kuan" zhiyin "kuan" lai wuyong'* ('China's Broadband: Life Is Not "Broad" because Being "Broad" Is of No Use'), *People's Daily*, 14 December 2001. Online. Available HTTP: <http://www.peopledaily.com.cn/GB/it/48/297/20011214/626895.html> (accessed 18 December 2001).
58 See *Zhongguo xinwen she* (China News Agency), *'Xinxichanyebu huyu Zhongguo yingdang jingti "xinxi geju" jumian'* ('MII Urges that China Should be Aware of the "Informational Compartmentalisation" Situation'), *People's Daily*, 28 October 2001.
59 Dai, *The Digital Revolution and Governance*, p. 120.
60 See R. Janda, 'Benchmarking a Chinese Offer on Telecommunications: Context and Comparisons', *International Journal of Communications Law and Policy*, 1999, issue 3. Online. Available HTTP: <http://www.ijclp.org> (accessed 14 February 2000).
61 J. Harding, 'Media: Still Maintaining Firm Control', *Financial Times*, 1 October 1999. Online. Available HTTP: <http://www.ft.com/ftsurveys/country/sc79b6.htm> (accessed 6 November 2000).
62 For more detailed discussions about Internet regulation in China, see Xiudian Dai, 'Chinese Politics of the Internet: Control and Anti-control', *Cambridge Review of International Affairs*, 2000, vol. 13, no. 2, pp. 181–94; and Chapter 3 by Gudrun Wacker in this volume.

2 Internet growth and the digital divide
Implications for spatial development

Karsten Giese

> I honestly believe that the new information economy has the potential, at home and around the world, to lift more people out of poverty more quickly than at any previous period in all of human history – and that tapping that potential is actually in our enlightened self-interest.
>
> (Bill Clinton, April 2000)[1]

If the official statistics are to be believed, the growth of the Internet in China is outshining developments in the United States and Europe and the majority of the international Web population will be Chinese speakers within the foreseeable future. The number of Chinese 'netizens' has already reached tens of millions since the Internet was liberated from the 'academic ghetto' in 1995, when the first connections began to be made outside universities and research institutions.[2] Such spectacular growth has made the Internet in China a focus of attention for international firms, human rights groups and development organisations alike. Great economic hopes are now set on the purchasing power of so many Chinese online. This 'network of networks' is also seen as a channel for promoting democratisation and the panacea for alleviating poverty and under-development.

The driving force behind such an explosion of activity and its consequent optimism is a government with a strong belief that the Internet and information technology (IT) are crucial factors for building international economic competitiveness and overcoming interregional development gaps at home. Yet this is not the first time that the PRC appears to have broken the mould of economic orthodoxy. While the current government's vision of the 'Third Technological Revolution' is, of course, not remotely comparable to the economic adventurism of Mao's Great Leap Forward, it will be argued below that a high degree of caution should be exercised when it comes to assessing the impact of China's digital 'leapfrog' if the true problems of the digital divide are not to be overlooked.

Problem of data

The first problem that must be confronted when assessing the expectations placed on the Internet in China is that they are based on decidedly shaky data. This is because, barring a few foreign enterprises specialising in market research with explicit interests of their own,[3] the sole provider of statistics on the subject is the semi-official China Internet Network Information Centre (CNNIC). It is according to this organisation that the number of Internet users grew from just 620,000 in October 1997 to 33.7 million by the end of 2001 (Figure 2.1), with the number of Websites reaching more than 277,000. It has also claimed that there were more than 60 million email accounts at the end of 2000, since when it stopped providing absolute numbers and simply claims that there is an average of 2.2 email accounts per user, which would give an approximate total of 74 million.[4]

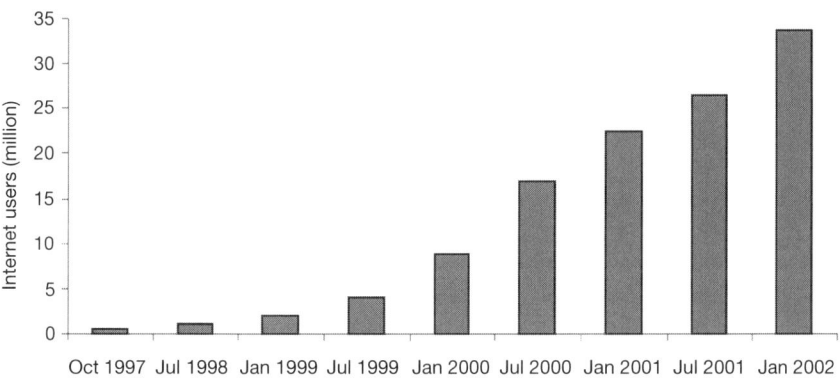

Figure 2.1 Internet growth in China (1997–2002)

Sources: CNNIC, *Zhongguo Internet fazhan zhuangkuang tongji baogao (1997/10–2001/7)*; CNNIC, *Zhongguo hulianwangluo fazhan zhuangkuang tongji baogao (2002/1)*.

While such figures look impressive, a closer inspection reveals that they have been arrived at through projections made on statistics garnered from methodologically questionable surveys. These are taken twice a year online,[5] with users guided to questionnaires through links on popular Chinese-language portals. The results are then supplemented by random interviews conducted among the general population and the use of various survey methods, the details of which are not specified.[6] The number taking part has fluctuated between 3,000 and 1.6 million, with Web users motivated to participate by a lottery until recently. According to CNNIC, the number of responses that were assessed as valid ranged between 1,800 and more than 570,000 (Table 2.1). It is on this uncertain methodological basis that the CNNIC has felt able to come up with figures such as 33.7 million Internet users at the end of 2001, even though China's own Ministry of Information

Table 2.1 CNNIC surveys on Internet development in China

Survey period	Netizens (million)	Survey methods	Data basis	
Unknown, qualifying date 31 Oct. 1997	0.62	Questionnaire in computer magazines; online questionnaire via 4 network providers and 8 Chinese ISPs.	Respondents: valid total:	N.A. 1,802
6–30 June 1998	1.17	Online questionnaire.	Respondents: valid total:	3,098 2,494
11–31 Dec. 1998	2.1	Online questionnaire via approx. 30 portals.	Respondents: valid total:	23,876 22,177
15–30 June 1999	4.0	Online questionnaire, linked via 'many popular portals'.	Respondents: valid total:	66,283 52,549
15–31 Dec. 1999	8.9	Online questionnaire, linked via 'many famous portals'.	Respondents: valid total:	363,538 202,432
22 May–30 June 2000	16.9	Online questionnaire, linked via 'many popular portals'; sample survey.	Respondents: valid: valid sample: valid total:	1,629,361 573,902 3,679 577,581

Table 2.1 continued

Survey period	Netizens (million)	Survey methods	Data basis	
End Nov.–31 Dec. 2000	22.5	Online questionnaire, linked via 'many famous portals'; sample survey, methods not specified; cross-checks by telephone; nationwide interviews.	Respondents: valid: valid sample: valid interviews: valid total:	34,695 26,667 62,620 6,000 95,287
Unknown–30 June 2001	26.5	Online questionnaire, linked via 'many famous portals'; telephone survey based on 'scientific sampling methods'.	Respondents: valid: valid sample: valid total:	144,083 78,342 4,828 83,170
Unknown–31 Dec. 2001	33.7	Online survey: questionnaire on CNNIC Website, linked via 'many famous portals';[a] telephone spot-checks among the general population owning fixed-line telephones more than six years, based on 'scientific samplings';[b] personal interviews of all students living on campuses.	Respondents: valid: valid sample: known valid total:	75,383 64,627 N.A 64,627 +

Sources: CNNIC, *Zhongguo Internet fazhan zhuangkuang tongji baogao* (1997/10–2001/7); CNNIC, *Zhongguo hulianwangluo fazhan zhuangkuang tongji baogao* (2002/1).

Notes

a According to the CNNIC report for January 2002 (*Zhongguo hulianwangluo fazhan zhuangkuang tongji baogao* (2002/1)), the questionnaire was directly linked via the portals www.sina.com.cn, www.163.com, www.sohu.com, www.gmw.com.cn as well as www.ced.com.cn.

b For details of sampling methods see CNNIC report for January 2002 (*Zhongguo hulianwangluo fazhan zhuangkuang tongji baogao* 2002/1).

Industry (MII), which holds administrative responsibility for the Internet economy, has produced the figure of 17.4 million for the same period.[7]

It must be admitted that measuring the number of Internet users is a worldwide problem, with published figures for any country to be regarded as estimations only. Perhaps the one thing that can be quantified with any degree of certainty, though, is the number of computers connected to the Internet. In China, this grew from 300,000 in October 1997 to more than 12.5 million by the end of 2001, according to the CNNIC.[8] However, As J. Eckblad has pointed out, even measuring the number of hosts is a poor guide to the number of actual users, because a single computer can serve any number of people.[9] And this uncertain method of measuring the number of users is rendered more opaque by the CNNIC because the organisation does not reveal the methods it uses for calculating the number of hosts it claims in its reports. The suspicion that the figures it provides are deducted from interviews, rather than electronic counts, is supported by the fact that the results indicate an obviously parallel trend in the number of Internet users and the number of computers connected to the Internet (Figure 2.2).[10]

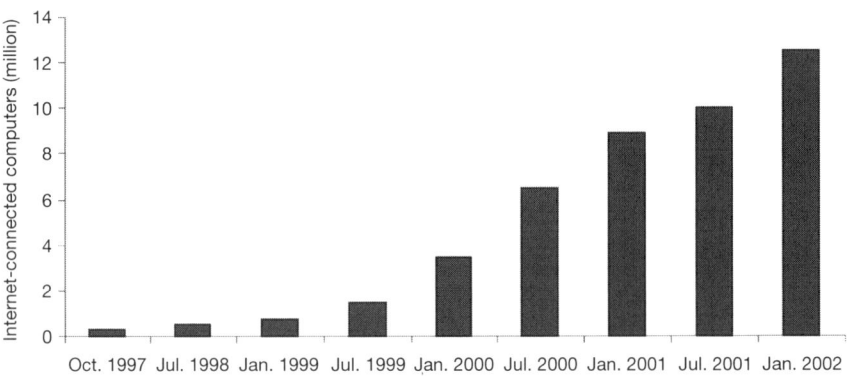

Figure 2.2 Internet hosts in Mainland China (1997–2002)
Sources: As in Figure 2.1.

In light of the above findings, it seems reasonable to conclude that the data provided by CNNIC should not be considered as a reliable basis for exact scientific analysis. Not only are the figures discredited by a lack of methodological transparency, but the organisation also appears to have altered its basic methods for collecting data several times already – a phenomenon typical of Chinese statistics in general.[11] It thus remains an open question as to how far and in what ways the CNNIC reports might be influenced and coloured by the political or economic interests of the Chinese government, of the IT industry or even of single actors in ministries and subordinated administrative units. At best, in the absence of alternative sources of data, the CNNIC reports might be considered to be reflections of rough trends, although even this assumption should be made with caution due to the way in which discontinuities arise from the practice of frequently changing the structure of the reports.[12]

Growth factors

Despite the unreliability of the available data on the development of the Internet in China, it is still possible to speculate that growth has been stimulated by a number of broad factors. The massive expansion of telephone networks since the mid-1980s, for example, can certainly be considered to be the first state-sponsored step within a strategy of increasing connectivity. Not having to replace an old system of copper cables has also been an undeniable advantage in wiring up the provinces through the construction of China's north–south and east–west backbones.[13]

It is also true that the country's political leadership evinces a strong commitment to the Internet. The priority given to the grand project of informatization is clear to see in recommendations for the Tenth Five-Year Plan (2001–5) published in October 2000 by the Central Committee of the CCP. This classifies the building of an information society as one of sixteen fields of strategic development for the industrialisation and modernisation of China. It is a vision that includes the widespread importation of information technology and widening the use of computers and data networks to exploit the full potential of information as a key resource. Plans exist to employ digital network technology for administrative and social work, as well as the management of commercial enterprises and the conduct of e-commerce. All of this, moreover, has ramifications for the education system, where the training of IT specialists and the spreading of the knowledge and skills necessary for survival in an information society is recommended. Finally, within the context of the 'Go West' programme for developing the western regions, all of this is supposed to help with the priority of lifting the economy in the least prosperous areas of China out of poverty.[14]

The proactive role of the state in projects such as 'Government Online', an initiative to use the Internet to promote administrative efficiency, may also have contributed to stronger growth. Because one aim of the government in promoting such projects is to take control of 'agenda setting'[15] by steering users towards content that it deems to be 'useful',[16] it has ended up officially sanctioning the need for Websites to enhance their contents and create appealing Web designs. As a result, recent years have seen the authorities positively encourage the emergence of an increasingly rich variety of Chinese-language content and visual and multimedia designs that meet international standards. The enormous expansion of content that has occurred since the end of the 1990s stands in stark contrast to the days when usage was limited to those with a good command of English and some rather sophisticated knowledge of computers (Figure 2.3).

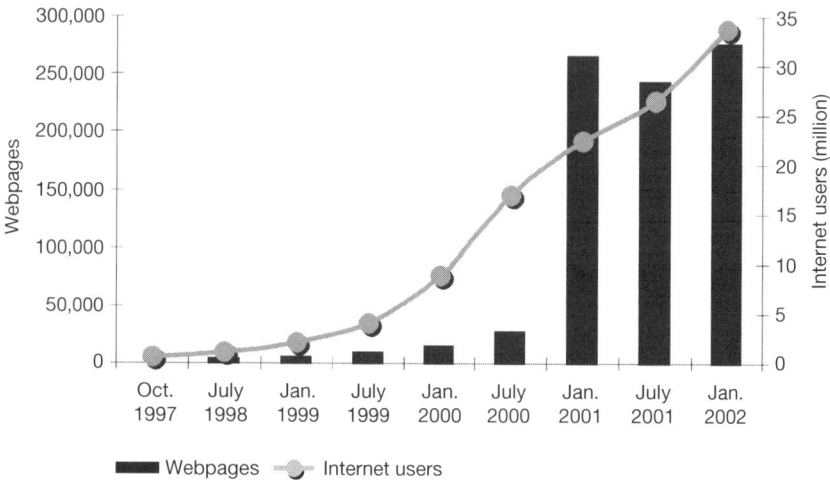

Figure 2.3 Chinese Internet users and WorldWideWeb contents (1997–2002)
Sources: As in Figure 2.1.

Finally, commercialisation of the Internet in line with the rules of the global 'New Economy' has also stimulated widespread usage. From this perspective, it is not surprising that the number of registered commercial domain names under '.com' reveals the same kind of acceleration in the Chinese Internet market that occurred in the industrialised countries, followed by a consolidation around the middle of 2001 (Figure 2.4). Overall, then, looking at the broad commercial and political factors that impact on the development of ICTs lends support to the view that China has witnessed an unbroken expansion of the Internet since the mid-1990s.

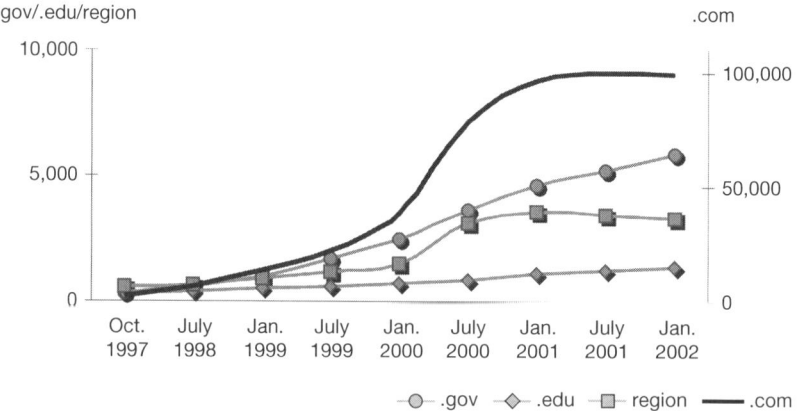

Figure 2.4 Second level domains under '.cn' (1997–2002)
Sources: As in Figure 2.1.

Causes for concern

Yet, despite these positive signs, a closer look at the CNNIC statistics reveals that euphoric predictions based on extrapolations from past trends should be treated with caution. First of all, the CNNIC studies published in January 2001 and July 2001 seem to corroborate international experience, which teaches that declining growth rates are to be expected after an initial phase of fast development. Although a revival of growth was indicated by the figures published in January 2002, only future surveys will confirm whether or not this means that China has bucked the international trend.

Perhaps more problematic is the limited size of China's Internet clientele base. While individuals with above-average incomes have made up the majority of Chinese 'netizens' in the past, the CNNIC surveys of July 2001 and January 2002 suggest that the largest group of potential users, namely urban professionals with an average income, has now been successfully tapped.[17] Similarly, while the high proportion of female Internet users (40 per cent) claimed by the CNNIC figures is a distribution characteristic of developed states, international experience again indicates that when such a point is reached it is usually followed by a clear decline in user growth rates. In other words, the possibility cannot be ruled out that China is already seeing the saturation of the primarily urban market for Internet access. If this is so, then even improved Web content may be unable to stimulate the positive impulses required for greater usage, especially in view of the high costs imposed by commercial providers,[18] restrictions on creating private Websites, and the already discernible consolidation of the sector.

State projects such as 'Government Online' may be unable to address this kind of problem. While pressure from the central government has led most administrative organs to have a presence on the Internet, the value of the information on the majority of such Websites has been marginal and rarely updated.[19] Yet, while the potential of the Internet to make citizens more familiar with governmental activities and to supply them with objective information has hardly been tapped, it is only during the past two years that there has been any real attempt to depart from the long-established practice of keeping information about governmental activities, political plans, the social situation, economic development, the environment or even certain administrative requirements confined within exclusive circles.[20] So far, although the government's strategy of using professionalism and public orientation to compete with the content offered by commercial providers has led to a visible increase in the volume and quality of publicly accessible information,[21] it is an uphill struggle to make China's bureaucratic culture compatible with the demands for transparency that characterise a modern information society.[22]

Bold predictions that there will be more Internet users in China than in the United States within the next couple of years[23] also start to look like wishful thinking when we realise that the user clientele revealed by the CNNIC is located almost exclusively in China's big cities. Although one study carried out in 22 of the largest cities in the year 2000 came to the surprising conclusion that 90 per cent of households were already subscribing to fixed-line telephone services, 60.4 per cent of families were using paging services and 42.7 per cent of individuals owned a mobile phone,[24] the official statistics cannot hide the fact that such growth has been a predominantly urban phenomenon and that regional comparisons paint a highly differentiated picture (Maps 2.1 and 2.2).[25]

Digital divide

Optimistic forecasts might also do well to take more account of China's technological lag behind the advanced industrial economies. The Chinese leadership is certainly concerned over the development of what has become known as the 'double gap' of development, namely falling behind the advanced economies in building both an industrial and an information society. The nature of this problem can be seen quite clearly in the structure of China's IT industry, which has successfully built an export-oriented sector while remaining entirely dependent on United States-sourced CPUs. In the software industry, despite government encouragement of the use of open-source operating systems like Red Flag Linux, Microsoft Windows has maintained its dominant market position.

When President Jiang Zemin delivered a speech to the millennium summit talks of the United Nations in the autumn of 2000, he thus acknowledged that scientific and technological innovation had become an important factor for creating wealth, but also expressed his fears over the continuing growth of the international digital divide and appealed to the industrialised states to share more scientific knowledge with the developing world.[26] In fact, by the mid-1990s, China had become the leading force among Asia Pacific Economic Cooperation (APEC) countries in efforts to promote the cultivation of human resources and the development of IT.[27]

The double gap impacts just as much on the digitial divide within domestic economies as it does on the gulf between states. The downside of China's technological backwardness for the broader domestic economy can be seen in the fact that high-tech industries account for only a tiny proportion of GDP, just 2 per cent, according to a *Xinhua News Agency* report at the end of 2000.[28] The proportion of the workforce in the electronics and telecommunications sectors is also small. Although this did grow from 2.5 per cent in 1996 to approximately 3.4 per cent by the end of 2000, even this moderate increase is largely attributable to the fact that the total number of industrial workers dropped by a third over the same period of time. In real terms, the IT and telecommunications sectors employed 1.38 million people in 2000, which is 250,000 less than in 1996.[29] There are also strong concerns amongst the leadership over the chronic lack of scientific and technical personnel, a phenomenon that is seen to be one of the greatest obstacles for

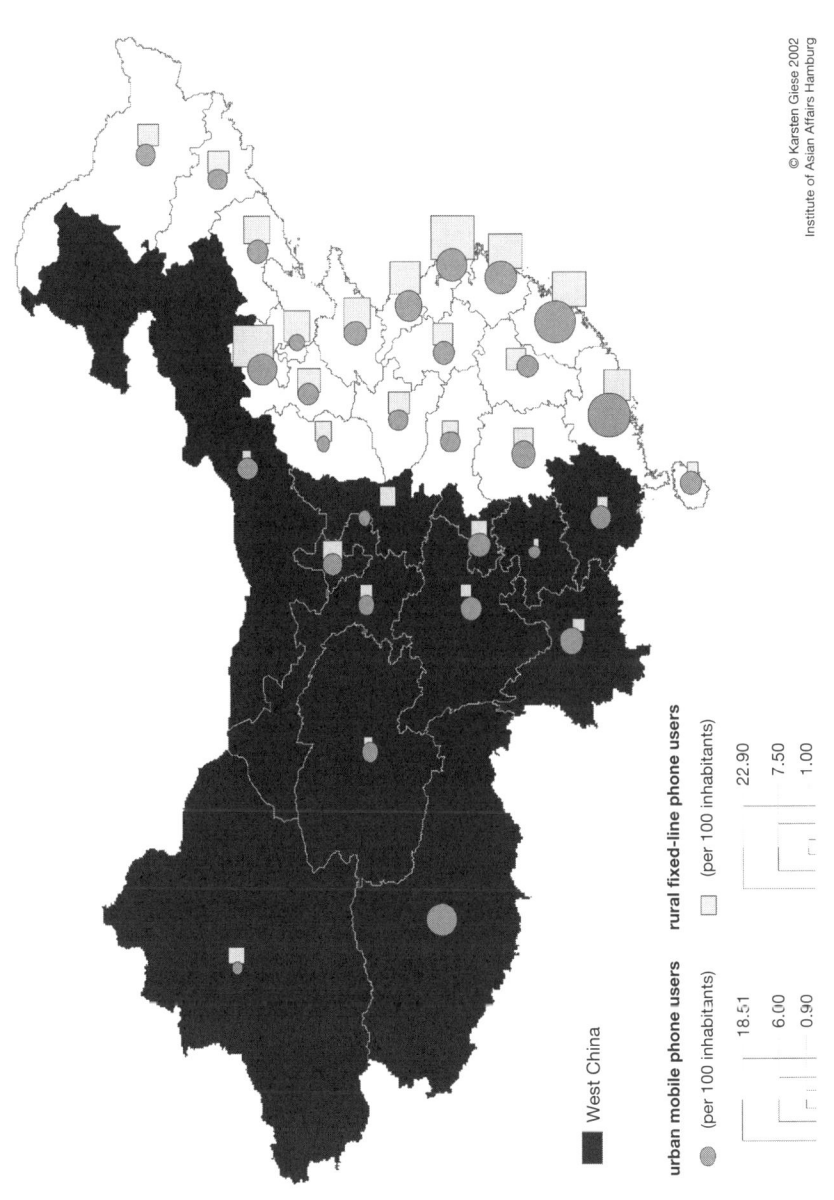

Map 2.1 Density of rural telephones and urban mobile phone subscribers in China (2000)
Data: National Bureau of Statistics, *China Statistical Yearbook 2001*.

international competitiveness. Although remarkable progress has been made in this area since the founding of the PRC, it is a sad fact that only a tiny fraction of the population of 1.3 billion is working in the high-technology sector today, with not even one employee out of 10,000 citizens employed in a scientific-technical profession.[30]

The domestic digital divide is thus characterised by a national situation in which a few high-tech islands stand isolated in what amounts to a vast technological wilderness. Even between cities it is possible to see intense disparities, correlating mainly with the decline in development and affluence from east to west. Beijing, for example, can boast that the industrial production value of its high-tech industries amounts to USD 2.47 bn, or 27.8 per cent of the capital's entire industrial output.[31] It is no big surprise that the CNNIC figures for December 2001 show that 9.8 per cent of Chinese Internet users were concentrated in the capital, as were 20.6 per cent of Websites and more than one-third of China's second level domain names under '.cn'.[32] Yet, according to a survey conducted by *ChinaOnline.com*, at the end of 2000, only 15 per cent of households in 22 large Chinese cities possessed computers with Internet access. Moreover, only 4.7 per cent had purchased Internet-compatible computers that year, amounting to a total of 867,000.[33] According to the National Bureau of Statistics, while the production of micro-computers had reportedly increased by more than 50 per cent to 6.72 million, only 9 out of 100 urban households possessed a PC.[34] Yet figures like these, low as they are to begin with, still disguise the seriousness of the real digital divide within China because there is a complete absence of statistics on the distribution of computers in rural areas. Looked at in this way, claims that there are tens of millions of Internet users in China become somewhat less impressive. In fact, even if we accept CNNIC's own projections of 33.7 million users in December 2001, this only accounts for 2.6 per cent of the population. The regional disparities are glaring, with Internet density ranging from 23.9 per cent in Beijing to a mere 0.57 per cent in the province of Guizhou (Figure 2.5).[35]

While it is true that Beijing's dominant position has been weakened over the years, it can be seen that this has not contributed to a more balanced distribution of the Internet across the country. Instead, it is cities like Shanghai and Tianjin, as well as urbanised centres like Guangzhou and special economic zones (SEZs) like Shenzhen and Zhuhai, that have profited. Moreover, although the overall share enjoyed by urban centres had shrunk from almost 54 per cent to 32 per cent by the end of 2001, the redistribution that occurred was still confined within the developed coastal regions of eastern China and had no real impact on the western regions. The fundamental disparities between regions have thus been left almost untouched (Map 2.3).

These regional disparities are reflected in the profiles of Internet users, who are still overwhelmingly located in cities like Beijing, Shanghai and Guangdong province. All of these are relatively affluent areas in terms of material wealth, and show high concentrations of institutions of higher

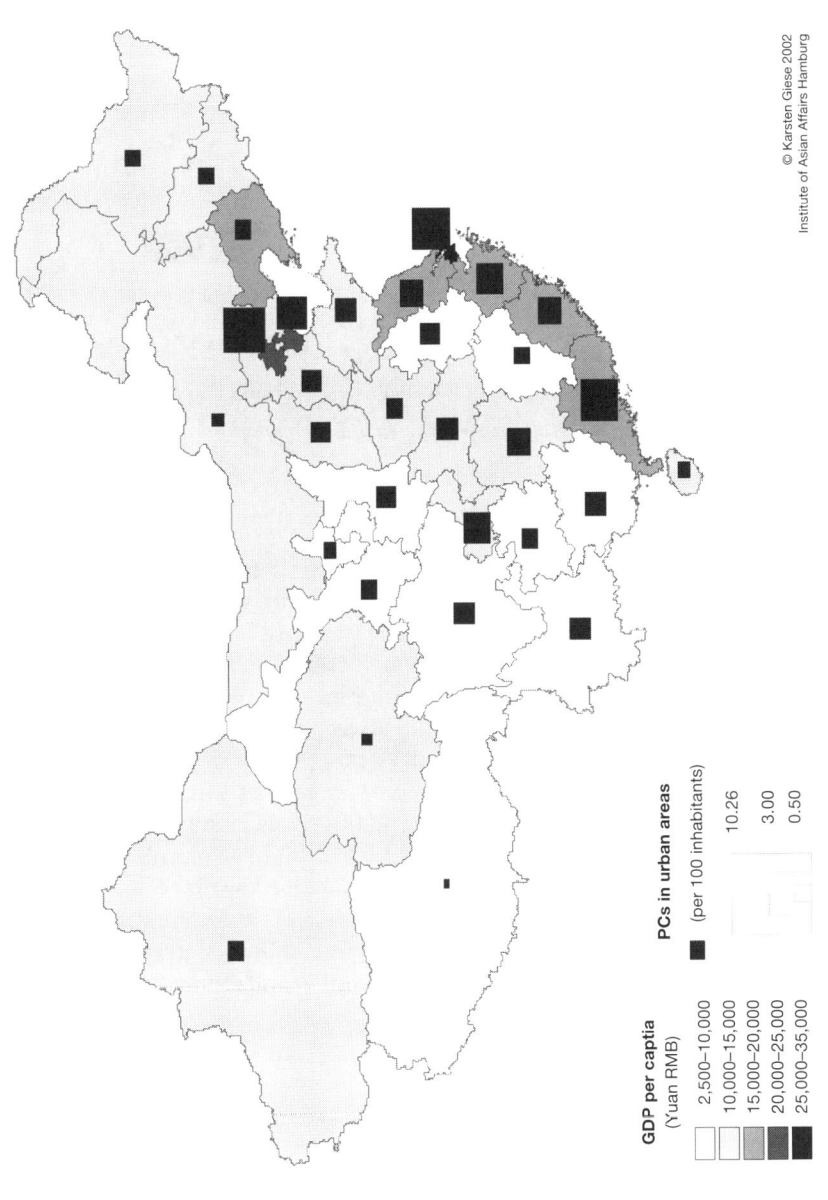

Map 2.2 Wealth distribution and urban computer ownership in China
Data: National Bureau of Statistics, *China Statistical Yearbook 2001*; calculations by the author.

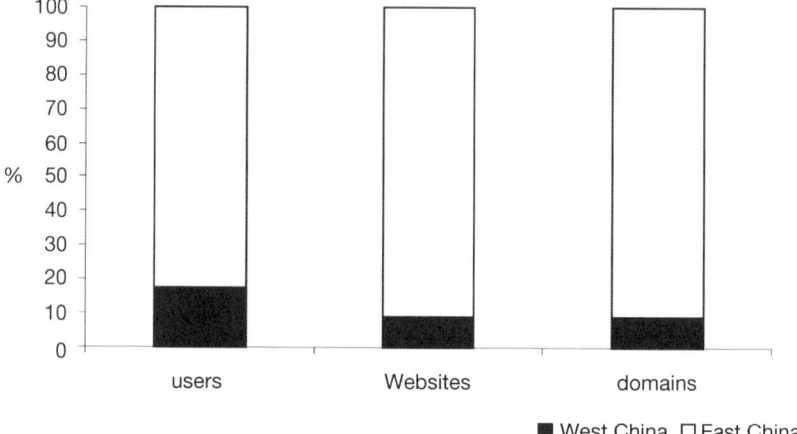

Figure 2.5 Geographic concentration of Internet users, Websites and domains* (2002)
Data: CNNIC, *Zhongguo hulianwangluo fazhan zhuangkuang tongji baogao (2002/1)*; calculations by the author.
Note
* -edu domains not included.

learning. The overall pattern is one of users from a relatively privileged strata of the population, dwelling in highly urbanised settings and concentrated in the prosperous eastern coastal regions. The majority is male (60 per cent), younger than 30 years (67.8 per cent), unmarried (56.9 per cent), with an above-average education (59 per cent with high-school diploma or higher degrees) and an above-average income of between RMB 500 and RMB 2,000 (50.6 per cent).[36] It should be acknowledged that positive developments have been claimed by the CNNIC, such as 23.6 per cent of Internet users now having a monthly income of less than RMB 500,[37] which puts access in reach of those with an average urban income of RMB 523 per month.[38] Yet any optimism derived from such figures needs to be qualified by the realisation that the average real income for the rural population stood at a mere RMB 187.[39] It can be concluded from these findings, therefore, that the Internet in China still represents a medium that is primarily for young and affluent city dwellers (Figure 2.6).

Why go west?

While insufficient infrastructure is still a problem for achieving connectivity in the countryside, and especially the western regions, high costs remain the root cause preventing rural dwellers from gaining access to the Internet. Although government intervention and increasing competition has led to price reductions, this remains a major reason of complaint for almost 40 per cent of users.[40] Particularly serious is the high price of Internet compatible computers,

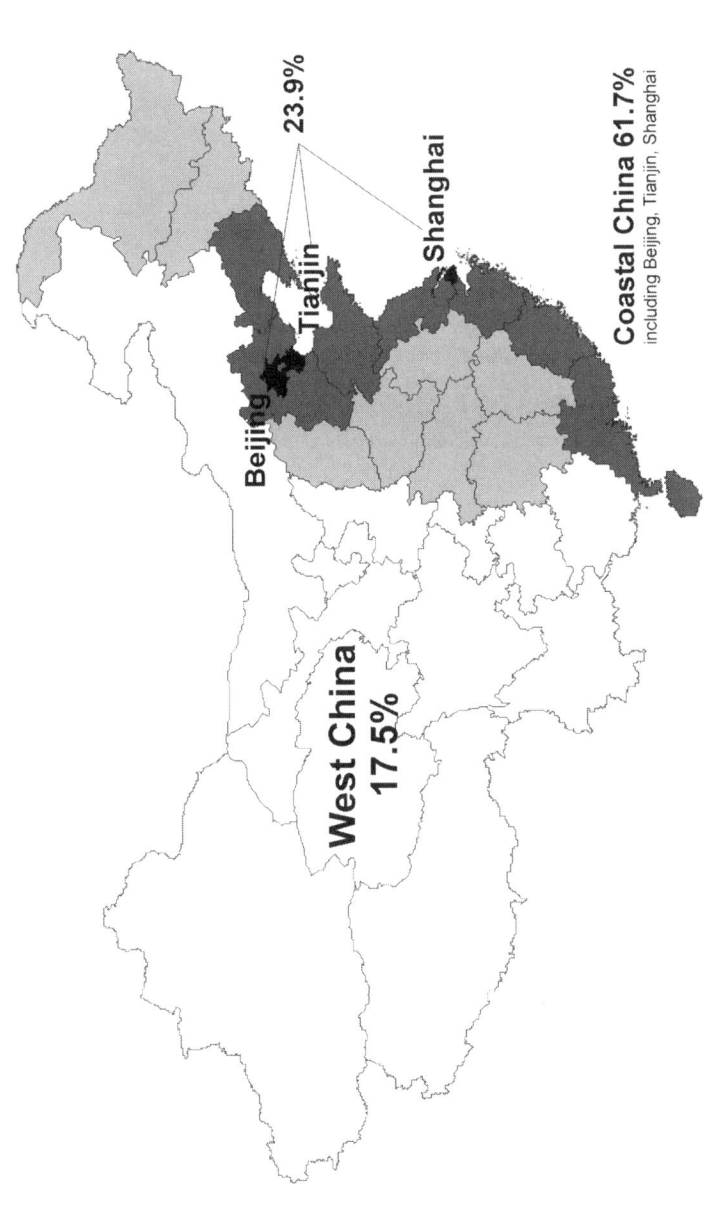

Map 2.3 Concentration of WorldWideWeb users in China (2001)
Sources: CNNIC, *Zhongguo Internet fazhan zhuangkuang tongji baogao* (2001/1); calculations by the author.

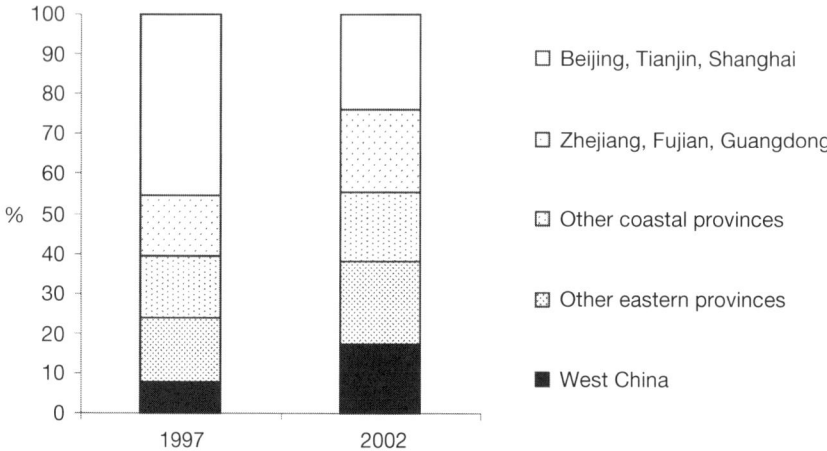

Figure 2.6 Prevalence of netizens in selected regions (1997 and 2002)
Sources: CNNIC, *Zhongguo Internet fazhan zhuangkuang tongji baogao (1997/10)*; *Zhongguo hulianwang fazhan zhuangkuangluo tongji baogao (2002/1)*; calculations by the author.

which remains well beyond the reach of the majority of the population.[41] The price of an average Pentium III PC in fact was roughly equal to the estimated per capita GDP of approximately USD 800 in the year 2000.[42]

Another aspect contributing to the digital gap between the countryside and the cities is the low level of education in rural areas. There are clear disparities between provinces and regions with respect to the percentage of illiterate persons and the distribution of higher educational qualifications.[43] Despite general improvements in the situation, even official reports lament that illiteracy rates are too high in the western regions. It is not unusual to find districts and towns with 20 per cent of the population not being able to read or write properly, and less than 5 per cent of children attending school.[44] A massive brain drain of the better educated from west to east only aggravates this educational problem still further,[45] making it perhaps the biggest obstacle of all for access to IT.[46] Although the central government has ambitious plans to improve this situation by means of 'e-learning', such initiatives may well fail because of the backwardness of the rural telephone network and financial restraints. Up to now, pilot projects have not even been accepted within target groups because of insufficient infrastructure and low levels of computerisation.[47] The commercial orientation of Websites also mitigates against the success of such projects. According to one survey, only 180 out of 1,100 e-commerce Websites offered remote electronic education.[48] Meanwhile, the state's broader education policy seems to do little to relieve the marked gap between east and west, geared as the central government is to making appropriations for educational tasks that show a clear preference for the highly developed coastal regions (Map 2.4).

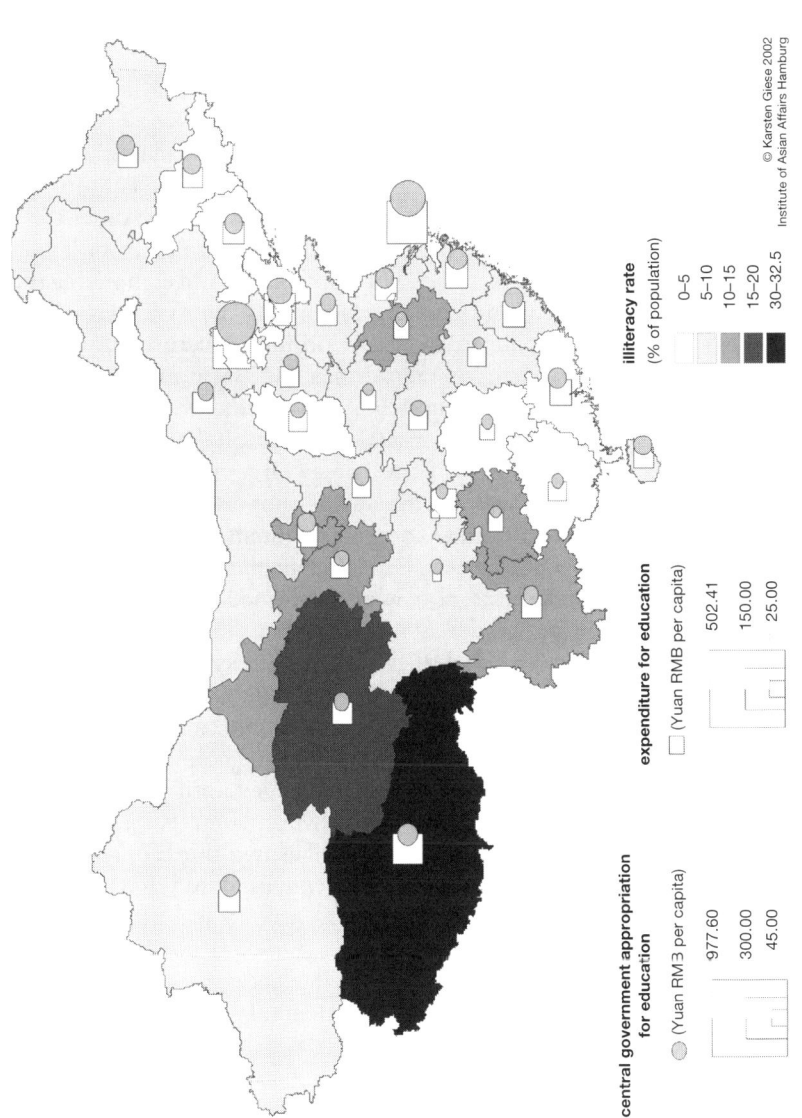

Map 2.4 Illiteracy rates and government funding for education (2000)

Sources: National Bureau of Statistics, *China Statistical Yearbook 2001*; calculations by the author.

This situation illustrates the need for the Chinese government to undertake real efforts on the education front. Yet even the provision of normal school education in the western regions remains highly questionable so long as sufficient funds cannot be found and strong enough political support is not forthcoming. Whether teachers and staff with adequate training and the will to work in underdeveloped areas can be found is also a big problem. If this is so for basic education, then the challenge that the government faces in providing the kind of education in ICTs that is necessary for the building of an 'information society' is daunting indeed.

It might be hoped that the situation in the poorest regions will be alleviated by the government's attempt to create a better economic environment through the national transfer of funds under the 'Go West' project, decided on in 1999 with the primary focus of improving infrastructure. Yet apart from laying approximately one million kilometres of fibre-optic cable and completing the installation of satellite telecommunications facilities,[49] the Tenth Five-Year Plan (2001–5) mainly aims at improving transport infrastructure. Moreover, while substantial amounts of government funds have been made available and foreign preferential loans are being channelled into the western regions,[50] it should not to be forgotten that enormous sums of capital are devoured by the largest single project of all, namely the Three Gorges Dam (Map 2.5).

Ultimately, therefore, the capacity of the state to support the creation of the necessary infrastructure for informatisation is very limited. Any catching up of the western regions with the economic development in the east of the country will thus have to depend in large part on investments made by enterprises and on their ability to recruit a qualified work-force. Yet the basic telecommunications providers have so far failed to give any real incentives for potential investors to go west. Although annual growth rates in the telecommunications sector averaged nearly 30 per cent in the 1990s, the only enterprise to profit from this was China Telecom, controlled by the state and holding a monopoly on fixed-line telephone services that was not challenged until January 2001. Despite market entry by the Ministry of Railways, some 95 per cent of subscribers to fixed-line telephone services as well as two-thirds of all mobile phone customers remained clients of China Telecom as of mid-2001.

Throughout the 1990s, China Telecom was able to use its monopoly to establish the nationwide telecommunication and data network with its backbones built on fibre-optic technology, a satellite network including thirty-eight major receiving stations, and the establishment of microwave networks in remote areas.[51] Now, however, China Telecom is unlikely to engage in this kind of network development because it has been listed on the stock market and faces national measures of control, forced price reductions for mobile phone services, a major streamlining of its operations and harsher competition as a result of WTO accession. Neither is it likely that the network infrastructure projects of its new competitor, Railcom, will significantly reach out into regions not already covered, since the strategy of its owner, the Ministry of Railways, is mainly to install lines along railway tracks.

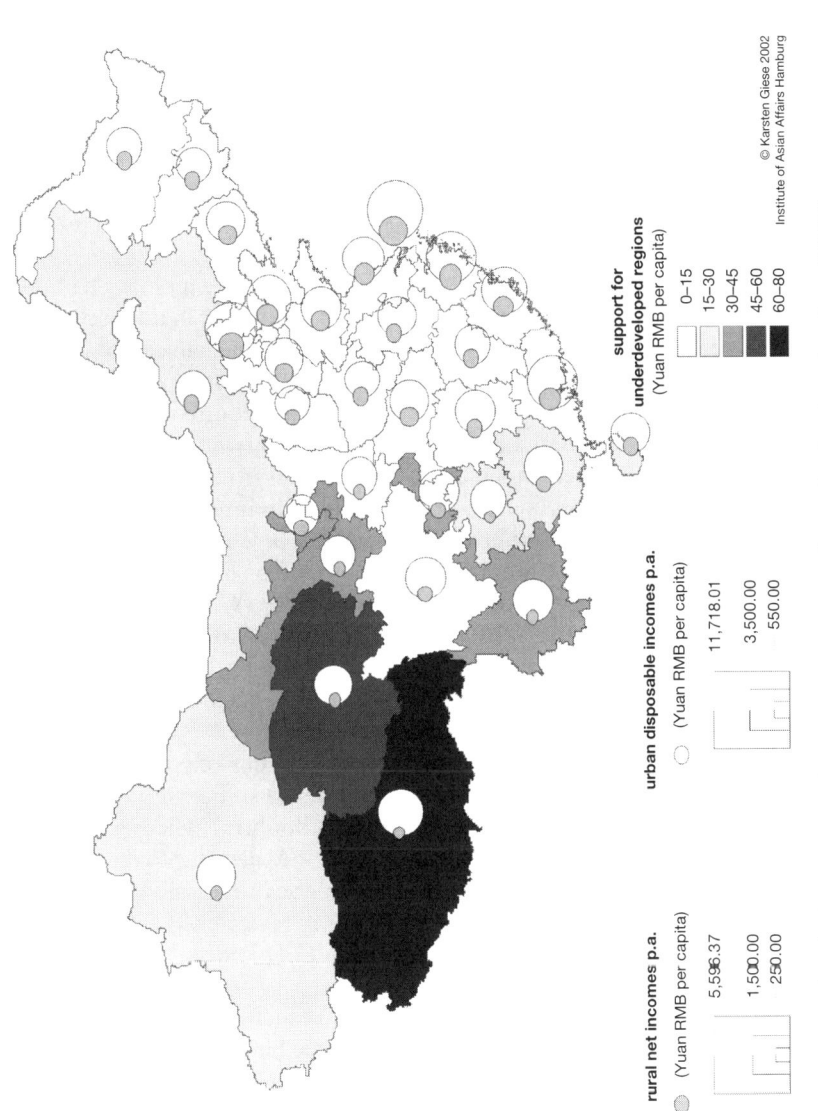

Map 2.5 Individual incomes and government support for underdeveloped regions (2000)
Data: National Bureau of Statistics, *China Statistical Yearbook 2001*; calculations by the author.

In view of this general situation, only limited stimuli can be expected from the big telecommunications providers for the national development of the Internet. The same applies to the other four Chinese competitors who have entered the Internet market and are capable of providing access to international data networks.[52] Although six network companies had gained the right to operate gateways between China and the international Internet by 2001, their domestic Chinese capacities still had to be leased from China Telecom.[53] When leasing network capacities accounts for 80 per cent of the operational costs for Chinese ISPs, in contrast to just 5 per cent in the United States, the distorted price structure functions as a strong disincentive for the expansion of services into new regions.[54] This situation is only made worse by the fact that ISPs shy away from investing in regions that show little promise of short-term profits, due to lack of purchasing power and low population densities.

No amelioration of the situation can be expected from the mobile telephone sector either. Despite enormous growth rates, the networks in this sector too have been concentrated along the east coast and in large urban centres, where market potential is relatively strong. As a result of China's topographical characteristics, expansion into the hinterland would require massive investment to build up new networks. So enterprises operating on a strictly commercial basis have been understandably reluctant to invest. As for mobile Internet access, experiences have so far been rather disappointing, making an expansion in this direction doubtful. In spring 2000, Chinese mobile phone operators as well as international equipment providers like Ericsson, Nokia and Motorola did promote Internet access via WAP (Wireless Application Protocol). However, because WAP failed to take off due to slow transmission speed and the lack of appropriate content, most providers had either stopped offering it altogether or had moved these services back to the fixed-line Internet by the end of 2000.[55]

As for the recruitment of qualified personnel, it is unlikely that education policies in the poorest areas will be able to establish a sufficient pool of well-trained specialists in the short term. It is also doubtful whether Chinese computer specialists, who are returning from Silicon Valley due to the bursting of the Internet bubble and the onset of the American recession to take up attractive job offers closer to home, will be willing to relocate to China's remote areas rather than settling down in the high-tech El Dorado of Beijing or the buzzing metropolis of Shanghai. Income and purchasing power, urban environment and the quality of life will thus be the decisive factors[56] when it comes to meeting the challenge of the digital divide between China's regions (Map 2.6).

Let them eat cyberspace

In sum, then, so far commercial concerns have not proved capable of assisting the central government's attempts to promote the use of IT, and especially the Internet, to help alleviate poverty and develop the rural economy. The grand project to overcome the digital divide has thus remained largely in the realm

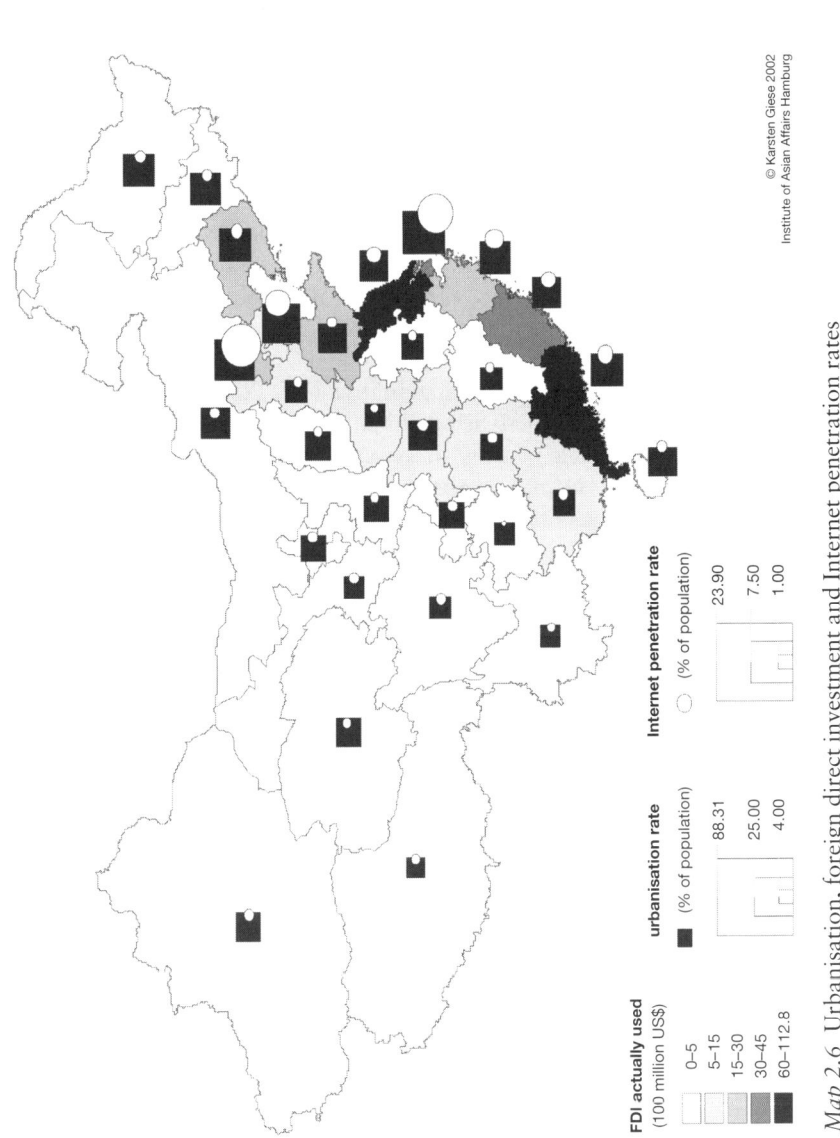

Map 2.6 Urbanisation, foreign direct investment and Internet penetration rates

Sources: National Bureau of Statistics, *China Statistical Yearbook 2001*; CNNIC, *Zhongguo Internet fazhan zhuangkuang tongji baogao (2001/1)*; calculations by the author.

of propaganda, due mainly to limited financial resources. The official media does sometimes carry reports of the successful deployment of the Internet in rural areas, such as news of a survey claiming how it had allowed over one million farmers in Hebei province to increase their profits by at least RMB 15 bn (about USD 1.8 bn) thanks to technological training, diversification of production and better sales information.[57] However, these stories tend to be of cases that are either very limited in terms of geography or time, or just too vague to make it possible to critically evaluate their broader implications.

Such stories certainly appeal to international development organisations, though.[58] The positive 'experiences' in Hebei province are confirmed by the United Nations Development Programme (UNDP) in China, which boldly states: 'It is strongly believed that improved access to technical and market information can substantially contribute to the reduction of recurring "hard core" poverty by increasing agricultural productivity and facilitating the commercialisation of agricultural products and to diversify the rural economy.'[59]

The UNDP is in fact running an imaginative USD 2.5 million project for taking Internet access to rural areas of China. At the centre of this is the idea of building 'community tele-centres' in five selected poor districts, which are supposed to broadcast information about market conditions, new agricultural technologies and methods of sustainable farming as well as social and health services. By establishing Internet terminals in local middle schools, pupils are supposed to act as information multipliers by reporting back to their families. In this way, it is hoped to reach most, if not all, households in the designated areas. During the first phase of the project, the information needs of the population in question will be identified in order to develop appropriate contents, and sustainability is to be achieved by training local personnel to manage and maintain the tele-centres (Map 2.7).[60]

Apart from faith, however, no convincing explanation is provided by the UNDP as to why it is better to use the Internet to inform and educate poor rural communities rather than more traditional means of communication. In fact, there remain well-grounded doubts about the feasibility of this whole approach. It is quite remarkable, for example, that only two of the locations selected for the pilot projects are actually located in the primary target regions for development chosen by the Chinese government, namely the 'west' (many of the places labelled as the 'west' are really located in what most people would see as central China). The UNDP planners are no doubt well aware of the fact that the distribution of information alone cannot improve the market position of rural producers if other requirements are not met, such as the provision of adequate transportation links.

In this respect, education is still probably the biggest problem. Although the UNDP addresses this issue with its measures to train staff in computer usage and maintenance, such an ambition has to be set against the lack of a positive educational ethos in the rural areas. Moreover, the severe shortage of computer-trained specialists even in the economic centres of the country

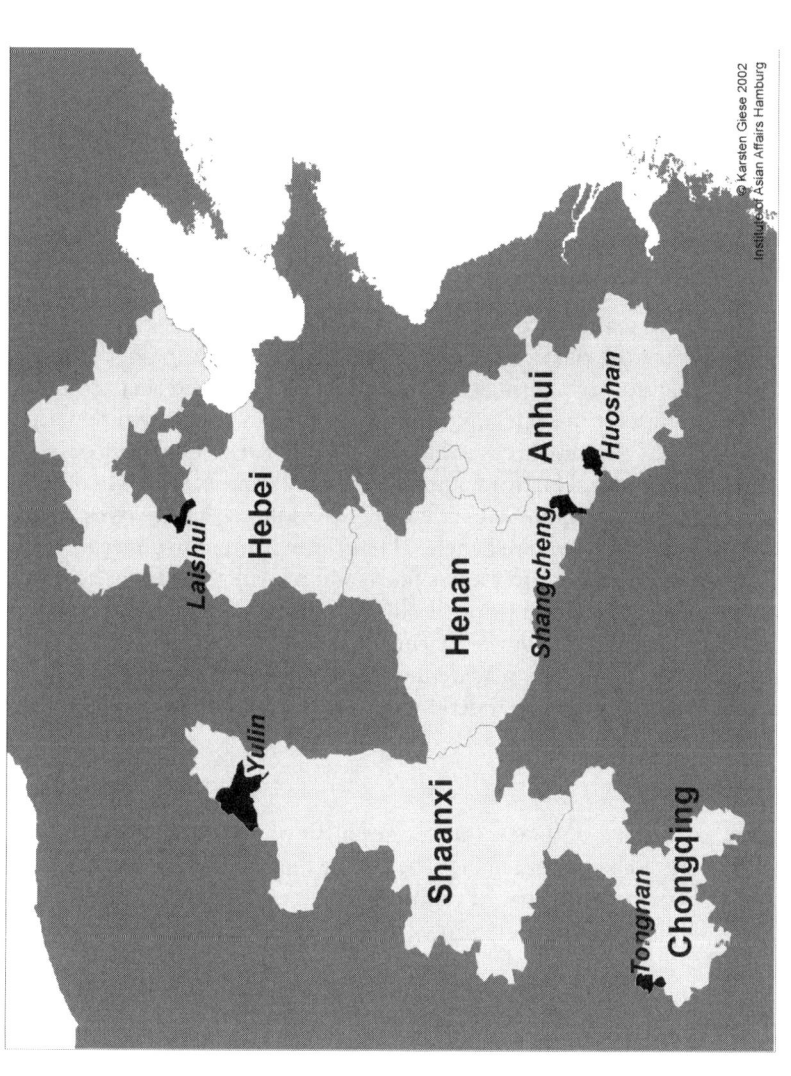

Map 2.7 UNDP Internet projects in China

Note
Provinces where UNDP Internet projects are located are named; the counties in which these projects are taking place are shaded black.

notwithstanding, enormous financial resources will be needed to train staff in this way. School staff will have to be adequately trained too, if the strategy of using pupils as multipliers is to work. Yet computer equipment is still insufficient even in universities,[61] and IT education is not regarded as even a supplement to the major subjects studied by trainee teachers. A lot of computers with Internet access will also be needed. Apart from the fact that no one knows how to pay for the necessary equipment, maintaining it will not be cheap at all. As one Chinese expert laments, the passage of Windows 3X, 95, 98 and 2000 in less than five years has already incurred great costs for those trying to keep up with the pace of change, making the Internet far from cheap for users and governments.[62]

One could always argue that there is no need to deploy state-of-the-art technology in rural areas. Yet this will merely consolidate the digital divide, leaving users in marginalised areas excluded from access to Internet resources not supported by the obsolete software and hardware that they are landed with. In the end, the inability to solve these problems goes back to the lack of determination on the part of the government to allocate the necessary funds. While the idea of structuring community access to the Internet around one networked computer in each community is not a bad one, it will remain meaningless without the resources necessary to support connection costs, training and maintenance. Hartford sums up the predicament well when she describes how wealthier regions are able to pour resources into networking infrastructure because they can see beneficial returns while central government funds fall far short of what poorer areas need, and providing Internet access is hardly at the top of the list of priorities in such places anyway. It can thus be expected that there will be some experiments and positive models that are highly visible, accompanied by a widening gap between the haves and the have-nots that is actually compounded by the access-to-technology gap.[63]

Conclusion

Despite some impressive statistics, the development of the Internet in China so far has been limited by two main factors. On the one hand, the government has decided to strive for improved administrative efficiency, control and planning by providing the necessary infrastructure for interconnecting its own institutions. On the other hand, growth has been driven by commercial interests generated by the Internet economy. Yet when it comes to bridging the digital divide, the government has lacked the political will and the financial resources needed to mobilise the Internet in a big way as part of its strategy to alleviate poverty and regional disparities. Under present conditions, however, commercial interests can hardly be expected to take up the slack when it comes to expanding connectivity to the peripheral and marginalised segments of the population.

Rather than expecting Internet access to speed up development in the backward areas, therefore, it is more realistic to expect its geographical distribution to follow the patterns of general economic development, improvements in income and educational levels, and the process of urbanisation. Yet even in the coastal provinces of eastern China there is sometimes little to be noticed in terms of a positive 'trickle down' effect from the most economically vibrant centres to the rural hinterland. Ultimately, therefore, the Chinese government faces a paradoxical situation when it comes to using the Internet as a tool for overcoming the digital divide: the level of connectivity that it wants to achieve in the western regions can only be achieved when a commercially attractive situation has already been created there.

It is hard to conceive of any way of reaching the government's goals for the western regions without falling back on commercial interests. During the period of 'reform and opening' the government has increasingly pinned its hopes on market forces for the injection of the finance necessary to achieve the objectives of its centrally planned development strategy. Yet when it comes to overcoming the geographical double gap of industrial development and the deployment of IT, market mechanisms appear to have acted more as obstacles than as the hoped-for remedy.

Ultimately, what the official statistics fail to tell us is that the Internet is unlikely to contribute to a more balanced regional development in China, because the Internet itself has developed in an extremely unbalanced way throughout the country. Moreover, if China follows international trends, then the rapid growth of the Internet that the country has witnessed so far can be expected to proceed at a markedly slower pace in years to come. Contrary to popular expectations, then, China's 'digital leap forward' could actually exacerbate the digital divide, rather than alleviate it.

Notes

1 Cited by M. Lacey, 'Clinton Uses High-Tech Show to Push Plan for Internet Parity', *New York Times*, 19 April 2000, p. 20.
2 ChinaOnline.com 'China's Internet Development Timeline'. Online. Available HTTP: http://www.chinaonline.com/issues/internet_policy/c9101571.asp (accessed 10 November 2000).
3 See, for example, China Internet Information Centre, 'How Many Users Are There in China', 8 February 2001. Online. Available HTTP: <http://www.chinaguide.org/english/7235.htm> (accessed 29 March 2001).
4 China Internet Network Information Centre (CNNIC), *Zhongguo Internet fazhan zhuangkuang tongji baogao (1997/10)* (*China Internet Development Statistics Report (1997/10)*). Online. Available HTTP: <http://www.cnnic.net.cn/develst/report1.shtml> (accessed 4 May 2001). CNNIC, *Zhongguo hulianwangluo fazhan zhuangkuang tongji baogao (2002/1)* (*China Internet Development Statistics Report (2002/01)*). Online. Available HTTP: <http://www.cnnic.net.cn/develst/2002-1/> (accessed 25 January 2002). Note that, according to the CNNIC, of the Websites under the Top Level Domain (TLD) '.cn' for 'China', only '.com', '.net' and '.org' were included in this figure, while '.edu' was excluded.
5 Methodological questions were discussed within the latest CNNIC report, *Zhongguo hulianwangluo fazhan zhuangkuang tongji baogao (2002/1)*, for the

first time. Unfortunately, this discussion was very limited and did not actually explain sufficiently how data were collected and statistics were aggregated.
6 Additionally, the definition of the term 'Internet user' is not transparent at all. Even internationally there is no standard for it. Since its July 2001 report, CNNIC has defined an Internet user as a person connected to the Internet for at least one hour per week. On the other hand, the same study reported an average of nine Internet hours per week. By the end of 2001, average Internet usage had declined to 8.5 hours per week, according to the CNNIC report for January 2002, *Zhongguo hulianwangluo fazhan zhuangkuang tongji baogao (2002/1)*.
7 Ministry of Information Industry (MII), *2001 nian 12 yue zhuyao tongxin zhibiao zhaiyaobiao* (*December 2001 Summary of Major Statistical Indicators for Telecommunications*). Online. Available HTTP: <http://www.mii.gov.cn/mii/hyzw/tongji/yb/tongjiziliao200112.htm> (accessed 7 February 2002).
8 See same sources as used for Figure 2.1.
9 J.G. Eckblad, 'Let them Eat Web. The Internet: Viagra or Sugar Pill for Treating Fledgling Economies?', *Student Technology Forum*. Online. Available HTTP: <http://www.ncsu.edu/connect/josh1.html> (accessed 25 April 2001).
10 If we include other sources, the suspicion of manipulated CNNIC reports grows. According to the *China Statistical Yearbook 2001* there were more then 14 million computers in urban households by the end of 2000 (calculations based on the following figures provided by the State Statistical Bureau: 9.72 computers per 100 urban households, 455.15 million non-military urban households and an average household size of 3.13 persons.) In light of statistics for past years one can assume that private computer ownership has a very short history in the PRC. Accordingly, it can be assumed that most of these 14 million computers were newer models connected to the Internet. In contrast, according to the CNNIC, there were only 8.9 million computers with Internet access for the same period of time. See National Bureau of Statistics, *China Statistical Yearbook 2001*, Beijing: China Statistics Press, 2001 (CD-Rom edition); calculations by the author; CNNIC, *Zhongguo Internet fazhan zhuangkuang tongji baogao (2001/1)*, (*China Internet Development Statistics Report (2001/1)*). Online. Available HTTP: <http://www.cnnic.net.cn/develst/cnnic200101.shtml> (accessed 4 May 2001).
11 Chinese officials have recently made some revealing statements on this topic. See James Kynge, 'Chinese Official "Confident" About Statistics', *Financial Times*, 28 February 2002, p. 6.
12 For example: in the mid-2001 survey report the CNNIC suddenly abstained from providing any data on geographical distribution of Internet users. In the latest study this information was included again. Therefore, we can only speculate about the reasons for the omission in July 2001. See *Zhongguo Internet fazhan zhuangkuang tongji baogao (2001/7)*, (*China Internet Development Statistics Report (2001/7)*). Online. Available HTTP: <http://www.cnnic.net.cn/develst/rep200107-1.shtml> (accessed 19 July 2001). Well-informed interviewees in China confidentially pointed to methodological shortcomings on the side of the CNNIC.
13 See Chapter 1 by Xiudian Dai in this volume, and also Xiudian Dai, *The Digital Revolution and Governance*, Aldershot: Ashgate, 2000; Zhang Junhua, 'Chinas steiniger Weg zum E-Government', in Gunger Schucher (ed.), *Asien und das Internet*, Hamburg: Institut für Asienkunde, 2002, pp. 97–111; M. Franda, *Launching into Cyberspace: Internet Development and Politics in Five World Regions*, London: Lynne Riener, 2002, pp. 188–92. Data of the National Bureau of Statistics, *China Statistical Yearbook 2001*, shows that Chinese telecommunication networks had a length of 2.2 million km in the year 2000, of which approximately 286,000 km were fibre-optic cable and 122,000 km microwave lines. By the end of 2001 the MII had reported a total length of 1.46 million

km for the national fibre-optic cable network. Such an enormous build-up within one year is surprising, to say the least. See Ministry of Information Industry (MII), *2001 nian 12 yue zhuyao tongxin zhibiao zhaiyaobiao, (December 2001 Summary of Major Statistical Indicators for Telecommunications).* Online. Available HTTP: <http://www.mii.gov.cn/mii/hyzw/tongji/yb/tongjiziliao200112.htm> (accessed 7 February 2002).
14 'Five-year Draft Plan Depicts Further Development', *China Daily*, 18 October 2000. Online. Available HTTP: <http://www.chinadaily.net/highlights/plan/index.html> (accessed 18 December 2000). An English version of the complete text of the outline of this plan delivered by Zhu Rongji at the Fourth Session of the Ninth National People's Congress on 5 March 2001, can be found in 'Premier's Report on Outline of New 5-Year Plan (Full Text)', *Xinhua News Agency*, 7 March 2001. See also Heike Holbig, 'Reformanlauf ins neue Jahrhundert – Offizielle und inoffizielle Agenda der 5. Plenartagung des XV. ZK', *China aktuell*, October 2000, pp. 1167–72; Dai Xiudian, 'Towards a Digital Economy with Chinese Characteristics?', paper presented at 'Development and Impact of the Internet in China' workshop, London School of Economics, 8–10 December 2000.
15 On Chinese central government plans for e-government, see Zhang Junhua, 'Chinas steiniger Weg zum E-Government', pp. 97–111.
16 On aspects of control, censorship, and self-censorship, see Chapter 3 by Gudrun Wacker in this volume.
17 CNNIC, *Zhongguo Internet fazhan zhuangkuang tongji baogao (2001/7)*, and *Zhongguo hulianwangluo fazhan zhuangkuang tongji baogao (2001/1)*.
18 Franda, *Launching into Cyberspace*, pp. 188–92.
19 It seems that quite a number of administrative Websites have not been uploaded properly. Several institutions also seem to be online on a pro-forma basis only, merely offering empty pages. On this, see 'Government Online: A Very Partial Success', 10 October 1999. Online. Available HTTP: <http://www.chinabiz.org/articles/it/991005.htm> (accessed 20 December 2000). One can get an impression of this project – at least as far as the central government level is concerned – via the Website: http://www.china.org.cn/search/gg/Government.htm. By February 2001 direct links to sixty-four institutions of the Chinese government had been established here.
20 Cf. 'PRC Internet: Cheaper, More Popular And More Chinese. An October 1998 report from U.S. Embassy, Beijing'. Online. Available HTTP: <http://usembassy-china.org.cn/english/sandt/Inetcaswb.htm> (accessed 20 November 2000).
21 On control and censorship issues in the Chinese Internet in this context, see also Chapter 3 by Gudrun Wacker in this volume; K. Giese, 'Big Brother mit rechtstaatlichem Anspruch. Gesetzliche Einschränkungen des Internet in der VR China' in B. Engels and O. Nielinger (eds), *Elektronischer Handel in Afrika, Asien, Lateinamerika und Nahost*, Hamburg: Deutsches Überseeinstitut, 2001, pp. 127–52 (Schriften des Deutschen Überseeinstituts, No. 50).
22 Governments have established Websites worldwide, but the level of development of those sites shows huge differences. Probably the most progressive development regarding government information provided through the Internet and the usage of this medium for communication with citizens can be witnessed in the USA. The publication of hearing protocols of former US president Bill Clinton concerning the Lewinsky affair is only one of the more exceptional examples for this development. For an overview of information and services of USA administrative bodies on the Internet, see http://www.savetz.com/yic/YIC09FI.html.
23 S. Thomas, 'Das Internet in China. Teil 1: Aufbau einer Informationsinfrastruktur', *China aktuell*, May 1999, pp. 500–10.

24 'Household Spending on Telecom Tools may Top US$774 Million in '01, survey says', 25 January 2001.Online. Available HTTP: <http://www.chinaonline.com/industry/telecom/currentnews/secure/b101012215.asp> (accessed 1 January 2001).
25 National Bureau of Statistics, *China Statistical Yearbook 2000*, Beijing: China Statistics Press, 2000 (CD-ROM edition).
26 *Xinhua News Agency*, 6 September 2000.
27 On the APEC plans and China's role, see China Secretariat for APEC, 'e-APEC Strategy Published', 22 October 2001. Online. Available HTTP: <http://www.apec-china.org.cn/APEC2001/20011022/928262.htm> (accessed 9 November 2001); China Secretariat for APEC, 'Shanghai Accord'. Online. Available HTTP: <http://www.apec-china.org.cn/APEC2001/20011021/927941.htm> (accessed 9 November 2001); *Xinhua News Agency*, 18 October 2001; K. Giese, 'China und die APEC', *China aktuell*, October 2001, pp. 1087–100.
28 *Xinhua News Agency*, 27 December 2000.
29 National Bureau of Statistics, *China Statistical Yearbook 2001*. Calculations by the author.
30 *Xinhua News Agency*, 27 December 2000.
31 *Xinhua News Agency*, 31 January 2001.
32 CNNIC, *Zhongguo hulianwangluo fazhan zhuangkuang tongji baogao (2002/1)*.
33 'Household Spending on Telecom Tools may Top US$774 million in '01, survey says'.
34 National Bureau of Statistics, *China Statistical Yearbook 2001*.
35 Calculations by the author based on data from CNNIC, *Zhongguo hulianwangluo fazhan zhuangkuang tongji baogao (2002/1)*, and National Bureau of Statistics, *China Statistical Yearbook 2001*.
36 According to the CNNIC report 12.3 per cent of users had no income of their own by the end of 2001, which is explained by the claim that 15 per cent of Internet users were under working age.
37 CNNIC, *Zhongguo hulianwangluo fazhan zhuangkuang tongji baogao (2002/1)*.
38 National Bureau of Statistics, *China Statistical Yearbook 2001*.
39 Ibid.
40 CNNIC, *Zhongguo hulianwangluo fazhan zhuangkuang tongji baogao (2002/1)*.
41 Thomas, 'Das Internet in der VR China. Teil 2: Nutzung und Inhalte von Online-Medien', *China aktuell*, June 1999, pp. 596–606; U. Schmid, china@wachstum.com, *NZZ Folio*, no. 2, February 2000, Online. Available HTTP http://www-x.nzz.ch/folio/archiv/2000/02/articles/schmid.html (accessed 20 April 2001).
42 *People's Daily*, 20 September 2000.
43 National Bureau of Statistics, *China Statistical Yearbook 2001*.
44 Yunnan, Sichuan, Ningxia, Qinghai as well as Tibet and Xinjiang are normally counted as Western Chinese regions. Unfortunately they were not named specifically. See *Xinhua News Agency*, 27 December 2000.
45 Ibid.; 'West Set to Reverse the Brain Drain', *China Daily*, 5 February 2001. Online. Available HTTP: <http://www.china.org.cn/english/7064.htm> (accessed 6 February 2001).
46 By mid-2001 a share of 5.4 per cent of active Chinese Internet users even complained about the Internet requiring too much specialised knowledge. See CNNIC, *Zhongguo Internet fazhan zhuangkuang tongji baogao (2000/7)* (*China Internet Development Statistics Report (2000/7)*). Online. Available HTTP: <http://www.cnnic.net.cn/develst/cnnic200007.shtml> (accessed 4 May 2001). No question related to this matter was included in the December 2001 CNNIC questionnaire. See CNNIC, *Zhongguo hulianwangluo fazhan zhuangkuang tongji baogao (2002/1)*.

47 T. Plafker, 'Tapping China's Potential. Logistical Barriers Slow Push for Online Learning', *International Herald Tribune*, 16 October 2000. Online. Available HTTP: <http://62.172.206.162/IHT/SR/101600/sr101600k.html> (accessed 10 January 2001).
48 Wang Xiangdong, 'Mobile Communication and Mobile Internet in China'. Online. Available HTTP: <http://www.telecomvisions.com/articles/pdf/china_mobile_internet.pdf> (accessed 7 February 2002).
49 Ma Zhongshi, 'Business Goes West; Regional Growth is a Major National Priority', *CHINA 2000*. Online. Available HTTP: <http://www.china2thou.com/003p2.htm> (accessed 29 August 2001).
50 Department of Foreign Affairs and International Trade, Canada, 'The China Business Collection: China's Western Development Strategy', *China Perspectives 2001*. Online. Available HTTP: <http://www.dfait-maeci.gc.ca/china/business/WesternDev-e.asp> (accessed 29 August 2001).
51 See Zixiang Tan, W. Foster, S. Goodman, 'China's State-Coordinated Internet Infrastructure', *Communications of the ACM*, June 1999, vol. 42, no. 6, p. 47.
52 See Chapter 1 by Xiudian Dai in this volume.
53 Since 1999 BBNet and ChinaNet run their own inner-Chinese data networks, but regarding size and expansion of China Telecom's network, one has to conclude that China Telecom is still the *de facto* monopolist in this business sector. See Franda, *Launching into Cyberspace*, pp. 188–92.
54 Ibid.
55 New efforts based on the very successful Japanese model of NTT's (Nippon Telegraph and Telephone Corporation) DoCoMo within those parts of the mobile networks with higher transmission speeds aim primarily at the urban population. Services comprise news, M-mail, local weather forecasts, M-banking, M-booking, education, job offers and travel information. On this, see Wang Xiangdong, 'Mobile Communication and Mobile Internet in China'.
56 The abolishing of the household registration system that has been announced could help to create positive incentives, if existing urban privileges are abandoned too. Without this, it will hardly be possible to convince qualified staff with an urban household registration in eastern Chinese metropolises to move to the western parts of the country.
57 'Internet Users to Exceed One Million in Hebei', *Xinhua News Agency*, 22 March 2001.
58 Estimates show, for example, that 80 per cent of current World Bank projects already include IT components. Cf. J.G. Eckblad, 'Let them Eat Web'.
59 UNDP in China, 'UNDP Launched Its Project of "Poverty Reduction through Access to Information, Communication and Technologies" in China', 20 February 2001. Online. Available HTTP: <http://unchina.org/undp/news/html/010220.htm> (accessed 4 April 2001.
60 UNDP in China, 'Project Brief – CPR/00/202', 19 February 2001. Online. Available HTTP: <http://unchina.org/undp/news/html/010220-1.html> (accessed 4 April 2001).
61 One should bear in mind that big schools in modern urban centres of China only have a maximum of one PC for a class of 40–50 pupils. The situation is no different at universities. See Wang Qiming, 'Opportunities and Challenges. A Case of Internet Development in China', last updated February 2001. Online. Available HTTP: <http://www.dse.de/ef/digital/wang-e.htm> (accessed 17 April 2001).
62 Ibid.
63 K. Hartford, 'Cyberspace with Chinese Characteristics', *Current History*, September 2000, vol. 99, no. 638, pp. 255–62. Online. Available HTTP: <http://www.pollycyber.com/pubs/ch/home.htm> (accessed 10 June 2001).

3 The Internet and censorship in China

Gudrun Wacker

The global nature of the Internet, the wide geographic distribution of its users and the diverse character of its contents leads many policy-makers to believe that activity in cyberspace is beyond the regulation and control of any single state. With specific reference to China, former United States President Bill Clinton once compared controlling the Internet with 'trying to nail Jello to the wall'.[1] China's own President, Jiang Zemin, drew attention to the dangers of spreading 'unhealthy' information and appealed to the international community to develop common mechanisms for 'safe information management' when he gave a speech at an international computer conference in Beijing in August 2000. Editorials in the *People's Daily* speak of 'hostile' forces at home and abroad trying to infiltrate the country via the Internet.[2]

In light of such apprehensions, it is somewhat ironic that the rapid growth of the Internet in China would not have been possible without the support of the country's political leaders, who stress the importance of ICTs and especially the Internet for future economic development and integration into the global economy.[3] The leadership has even gone so far as to mobilise the Internet for political purposes through projects such as 'Government Online', initiated in 1999 to enhance the presence of ministries, administrative units and local government in cyberspace, furthering transparency and accountability by making more information accessible to citizens, and fighting corruption and fraud.[4] The Internet portal Netease, which operates in Chinese and English, won several awards for an advertising campaign that it ran on Chinese television in the autumn of 2000 under the slogan 'Power to the people'.[5]

This chapter will attempt to unravel this apparent paradox of a Chinese party-state that encourages the spread of the Internet on the one hand, while believing that it can monitor and censor those aspects of activity in cyberspace that it sees as destabilising, dangerous or 'unhealthy' on the other. To do so, it is necessary to look at the ways in which the party-state mobilises its own resources and co-opts the support of voluntary or involuntary collaborators to make the final outcome of the presumed battle between ICTs and the authoritarian state less predictable than Bill Clinton's comparison with Jello suggests. If such efforts can be seen to be in any way successful, then the belief that electronic communication over the Internet cannot be subjected to political control has to be called into question.

Regulation of the Internet

Regulation of the Internet is an issue that has been widely discussed in authoritarian and liberal-democratic states alike, centring on the two questions of whether it is necessary on the one hand and technically possible on the other. While there do exist radical proponents of 'digital libertarianism' who advocate unrestrained freedom of activity on the Internet,[6] it would be hard to find any government in the world that does not see the necessity of regulating certain aspects of electronic communication. While the emphasis does of course differ, all governments accept the need to regulate when ICTs threaten their traditional role of maintaining state sovereignty by preserving a tax base in the face of cross-border e-commerce, ensuring the security of sensitive data, and preventing cyber crime. Even liberal-democratic states tend to accept the need for control when it comes to issues like the dissemination of (child) pornography, racism, the instigation of violence, rightist extremism, and hate speech. It is hardly surprising, then, that laws to regulate activity in cyberspace have been drafted and passed in just about every country. Needless to say, the terrorist attacks on New York and Washington that occurred on 11 September 2001 have given a new urgency to the debate on 'civil liberties versus national security' throughout the world.[7]

Whether such regulation is technically effective, however, is another question, especially given the changing nature of the Internet under the impact of commercialisation. Although commercialisation does lead to ever more complex architecture and wider connectivity on the one hand, it also creates new possibilities for monitoring the activities of users and revealing their identity on the other. So-called 'geo-location programmes', for example, make it possible to locate users geographically by linking IP-addresses to countries, cities and postcodes.[8] The motives behind such developments are complex, including the growing need to be able to enforce the laws of a particular jurisdiction, target advertising, or ensure that a Website pops up in the right language.[9]

In light of the changes in Internet technology under the impact of commercialisation, a number of observers have begun to develop theories that cast doubt on the assertion that there is something inherent in the nature of the Internet that puts it beyond the control of the state. Lawrence Lessig,[10] for example, identifies four elements that explain the possibility of shaping behaviour in cyberspace, namely regulations, social norms, the market and the architecture of the Internet itself. It is the nature of the last of these factors, the combination of hardware and software that Lessig calls 'code', that is especially important to grasp if we are to understand the degree of political choice available to states when it comes to controlling the Internet. As Lessig explains, when you go into cyberspace you find some places where you need to enter a password to gain access (online services such as AOL, for example), and others where you do not need to be identified. Sometimes the transactions that you enter into leave traces that link them back to you. In some places you can have privacy through the use of encryption, while in others this is not an option. All such features are set by the code writers, and can be used to constrain some behaviour by making other behaviour possible or impossible. It is thus the code that embeds certain values or makes certain values impossible. In this sense, code is a kind of regulation, in just the same way that the architectures of real-space codes are a kind of regulation.[11]

It is important to note that Lessig does not deny that there will always be ways to circumvent the constraints imposed by architecture. His point, however, is that we cannot conclude that *effective* control is impossible only because *complete* control is not, any more than the fact that a particular lock can be picked or broken does not prove the total uselessness of locks in general.[12] To find out how cyberspace is regulated, therefore, we need to discover how the code regulates, who the code writers are, and who controls the code writers.[13] From this, it follows that any investigation into the nature of Internet control must extend to the ways in which governments are able to *indirectly* regulate the Internet by *directly* regulating intermediary actors like Internet service providers (ISPs) and Internet content providers (ICPs).[14]

James Boyle, too,[15] has drawn attention to the ways in which states can regulate the Internet through a combination of privatised methods of enforcement and state-sponsored technologies by appealing to the theories of Michel Foucault concerning the subtle private, informal and material forms of coercion that are enforced through 'surveillance' and 'discipline'.[16] Such a vision is illustrated by the concept of the Panopticon, a prison in which every cell has a window facing a central tower from which an unseen warden just might be watching what any individual prisoner does at any time. The prisoner thus has to act *as if* he is under constant surveillance all the time, even though such surveillance is not physically possible for the single warden. By applying such a model to the kinds of technologies that are being built into the Internet, Boyle argues that effective censorship has indeed become possible.

Take common measures adopted in liberal-democratic states to enable individual Internet users or ISPs to shield people from undesirable content, such as software used to stop children from gaining access to pornographic Websites. SurfWatch, CyberPatrol, NetNanny or CyberSitter, for example, all contain lists with Internet addresses of 'forbidden' Websites and filters to block access when specific words or phrases are detected. Such activity, moreover, is encrypted and thus invisible to the user. Similarly, under the system of self-description known as 'Platform for Internet content selection' (PICS), suppliers can build certain information into Websites containing details such as the age group for which specific content is recommended. However, as Boyle argues, the very fact that such systems are claimed to be 'value neutral' means that their political impact will depend on who is using them, because 'The third party filtering site could be the Christian Coalition, the National Organization for Women or the Society for Protecting the Manifest Truths of Zoroastrianism'.[17]

It is important to realise that such technical solutions enable states to enlist private or commercial parties to carry out tasks that the state itself is either not allowed to carry out for constitutional reasons, or is unable to carry out due to its limited capacities. In fact, commercial ISPs and ICPs are far more susceptible to pressures from states in this respect than are individual Internet users, because they are not so able to operate independently of a geographical base or a real identity. When, for example, ISPs are made legally liable for any copyright infringements committed by their customers, they have to respond by erecting 'digital fences' to prevent unauthorised copying. Similarly, if encryption technology that is developed to enable activities such as online banking might also be used to enable dissidents and human rights organisations to communicate free from state surveillance, states can still normally justify legislation allowing them to gain access to electronic communications on the grounds that encryption can also be used to organise criminal activities.[18]

The kinds of developments indicated by writers like Lessig and Boyle should, then, be taken as warnings that the exercise of state power and control in cyberspace is not always easy to recognise. The methods used by states often appear to be 'natural' and integral parts of electronic communication media themselves, which they do not necessarily have to be. It is certainly premature to write off the ability of states to bring the Internet under control, when efforts to regulate and control cyberspace have been going on in practically every country, and governments have found allies in the form of commercial enterprises developing appropriate technologies. As Qiu points out, from a global perspective, the kinds of measures imposed by the Chinese government to achieve 'virtual censorship' are not so special, reflecting as they do 'the emerging attempts of legislatures, governments and various administrative organs worldwide to incorporate the cyberspace into their sphere of jurisdiction'.[19] In this light, it is important that we take into account the complex relationships that exist between the kinds of factors of control listed by Lessig and Boyle in any attempt to understand how the Chinese state censors the Internet.

National regulations

While regulation of the Internet in China is partly based on laws that pre-dated its existence, a series of specific regulations has also been introduced,[20] probably encouraged by the approach of WTO accession. Despite the ongoing streamlining of the state apparatus, formulating this mass of regulations has involved a confusing number of ministries and administrative units.[21] Many are intended to benefit Internet users, such as measures protecting consumers by governing online trading in pharmaceuticals and online educational services, or upholding intellectual property rights and individual privacy. In this respect, regulation in China is not so different from that found in other countries, and such provisions cannot be exclusively interpreted as efforts to stamp out political dissent.

More directly related to questions of control and censorship, however, are various provisions that were included in a raft of regulations that was introduced in the year 2000 to govern telecommunications and the publication of news and electronic information on the Internet. These will be listed below under the categories of 'forbidden contents', 'restrictions on the distribution of news', 'licences', 'storage of user data', 'surveillance', 'judicial liability', and 'penalties'.

1   *Forbidden contents*   Since the list of contents that are banned from distribution or electronic publication gives a full overview of the government's targets in controlling cyberspace, it is worth citing this in its entirety:

Any information that involves the following is forbidden:

(1) Contradicts the principles defined in the constitution [of the PRC].
(2) Endangers national security, discloses state secrets, subverts the government, destroys the unity of the country.
(3) Damages the honour and the interests of the State.
(4) Instigates ethnic hatred or ethnic discrimination, destroys the unity of [China's] nationalities.
(5) Has negative effects on the State's policy on religion, propagates evil cults or feudal superstition.
(6) Disseminates rumours, disturbs social order, undermines social stability.
(7) Spreads lewdness, pornography, gambling, violence, murder, terror or instigates crime.
(8) Offends or defames other people, infringes upon the rights and interests of other people.
(9) Other contents that are forbidden by law or administrative regulations.[22]

It should be noted that this catalogue of prohibited contents is not new. In a slightly more general form, it can be found in the regulations for the publishing industry issued in 1997.[23] Since contents banned on the Internet are nearly identical with those prohibited in other media, it can be concluded that the Internet and electronic information services are basically treated like other forms of publication.

2   *Restrictions on the distribution of news*   In principle, the regulations prohibit the distribution of news through the Internet, unless this news has either been published on the Internet by the official state-owned media or the news departments of state institutions themselves, or has already

been published by authorised media in another form. If a pure Internet portal wants to publish news via the Internet, it not only has to fulfil specific requirements (such as a professional editorial board, sufficient financial means and technical equipment), but also has to enter into a formal co-operation with one of the state-authorised media. The co-operation agreements have to be filed with the authorities in charge. The source and date of any published news has to be cited in each case. As for the question of whether Internet portals can publish news and articles by their own journalists, contradictory statements were made prior to the regulations and there is still a lack of clarity over this.[24] With respect to links to foreign news sites or the publication of news taken from foreign media – some portals had signed agreements with foreign information services like Dow Jones and others – the regulations stipulate (§ 14) that the prior consent of the Information Office of the State Council is necessary.

The way in which these rules try to kill two birds with one stone should be noted. On the one hand, they have a political or ideological purpose of containing and directing the proliferation of news material into relatively manageable channels. On the other hand, they aim at securing the economic interests of the official media *vis-à-vis* pure Internet portals operating in Chinese, such as Sina, Netease or Sohu, which have become very popular due to their early Web presence and timely and attractively presented news services. Faced by such competition, 'traditional' state-owned media such as *People's Daily* and the *Xinhua (New China) News Agency* formed an interest group as early as 1999. Pointing to the risks created by allowing the Internet portals too much freedom, they protested and appealed for official support from the state to improve their own Web presence. Fighting news piracy was one of the main issues on their agenda.[25] In order to strengthen the competitive position of such official organs, the State Council granted special funds (USD 121 million) to the most influential among them for launching and improving their own Internet presence.[26] One of the tasks of the 'Management Office for Internet Information' (*Guowuyuan Xinwen Bangongshi Wangluo Xinwen Guanliju*) established in April 2000 under the Information Office of the State Council, is to improve the online presence of such media.[27] There is thus a tight relationship between the economic and ideological dimensions of the regulations, since strengthening the presence of the state-owned media also helps to ensure that the supply of Internet news content is in line with what the state wants.

3   *Licences* Comprehensive and detailed rules on licences force providers of Internet services to apply for a range of special permits. These have to be obtained from different authorities, with a separate licence necessary for each category of service. The official registration numbers have to be displayed clearly on the Website in question. It is not permitted to expand

the business activity covered by such a licence without prior consent. All Internet cafés have to register with the local Public Security Bureau. Moreover, by requiring businesses to meet certain preconditions in terms of personnel, financial and technical equipment, market entry costs have been raised in ways that leave smaller independent Internet companies that have no official backing in a disadvantageous position compared to the big players. In this way, the 'market' (according to Lessig's categorisation) is brought into play in a legal way as a restrictive factor.

4   *Storage of user data* ISPs are required to store all user data. This does not only include the registration or customer number of the user, but also which telephone number is used for logging on, which Web addresses or domains are visited during the session and for how long. The data has to be stored for sixty days and be disclosed to the authorities on request. Providers of electronic information services, such as ICPs offering BBS (bulletin board service), chat room and discussion forum services, have to store all contributions published on the Internet, including time of publication and Web address or domain name, and keep such data for sixty days. Postings from users that violate any of the rules on banned contents have to be deleted from the Internet immediately. At the same time, however, they have to be locally stored and reported back to the authorities.[28]

5   *Surveillance* Reports in the foreign media have claimed that regulations introduced in January 2002 are the most intrusive so far for requiring ISPs and ICPs to screen e-mails.[29] However, it should be noted that the wording of the passages quoted above could also be interpreted as implying that providers are *not* allowed to screen the non-public communications of their customers, such as e-mail.

6   *Judicial liability* Providers of electronic information services have to notify their users or customers of the legal responsibilities that apply when posting and uploading information or contributions.[30] For the published contents, final responsibility rests with the respective author.[31] This is in contrast to Singapore, where it is ICPs that are made liable for all contents published within their business sphere.[32] The service providers can be held liable in China, though, if they operate without the required business licences, fail to meet their obligations with respect to the storage of data and notification of the authorities,[33] and also if they fail to protect personal data by passing them on without the prior consent of the respective user (although there are many exceptions to this stipulated by law, of course).

7   *Penalties* With respect to ISPs and ICPs the regulations provide for reprimand, rectification and fines, and even for the closure of Websites

in serious cases. If, for example, the licence number is not visible on a Website, a fine of RMB 5,000–50,000 can be imposed. If the obligations to delete contents violating the regulations and to notify the authorities of the violation are not met, the penalty can be revocation of the business licence.[34] As for the penal consequences of violations by ISPs, ICPs and individual users, the regulations refer to the 'relevant laws'.[35]

In sum, we can see that this raft of recent regulations constitutes an effort by the Chinese government to make a more proactive Internet policy and to provide itself with better instruments for influencing activities in cyberspace. This comprises not only content restrictions on the 'ideological' level, but also administrative and economic requirements which strengthen the official media as well as consolidating the advantages of bigger and financially well-equipped Internet enterprises.[36] Through these measures, and by offering more contents through projects such as 'Government Online', the CCP and the government hope to be able actively to set the agenda for the Internet in China.

Influencing norms of behaviour

When looking at the measures taken to enforce regulations by the Chinese state in the realm of the Internet, the number of 'political' cases (as against cases such as breaking into banking networks) with penal consequences has been rather modest so far, with between 10 and 20 people arrested and sued as of mid-2001.[37] The first of these to be made public was the case of Shanghai businessman Lin Hai, who in early 1998 sold 30,000 e-mail addresses in China to a New York-based organisation that distributes the pro-democratic newsletter *VIP Reference* (*Dacankao*) via e-mail in China. The involuntary recipients of this service included some high-ranking party officials. Lin, who claimed that his actions were strictly profit-driven, received a 2-year sentence, but was released early.[38] Another high-profile case was that of Huang Qi, who established the Website 'www.6-4tianwang.com' for relatives of missing persons, for which he had even received official praise after clearing up some cases of abduction. However, when contributions to the site began to commemorate the events that occurred in Tiananmen Square on 3–4 June 1989, the authorities closed down the Website and arrested Huang, who was accused of attempting 'to subvert the government and destroy national unity'.[39] Other known cases include that of a teacher in Nanchong who owned an Internet café and was arrested in August 2000 in relation to BBS-contributions criticising the Communist Party.[40] The organisers of the officially closed Website of the 'New Culture Forum' (www.xinwenming.net) were reportedly being sought by the authorities in the autumn of 2000.[41]

The Falungong spiritual movement has become the most prominent example illustrating the subversive potential of using the Internet from the perspective of the Chinese government. It has been claimed that electronic means of communication like e-mail played a central role when its members secretly planned and organised a mass demonstration in April 1999, right in front of the CCP's headquarters at Zhongnanhai, Beijing, which seems to have caught the Chinese leadership completely by surprise. The group, whose spiritual leader lives in the United States, certainly propagates its ideas on a number of Websites outside China. When the movement was declared illegal after the demonstration in Beijing, a fairly orthodox campaign to criticise it was accompanied by the transformation of cyberspace into something of an electronic battlefield as the authorities sought to paralyse servers housing the group's Websites through measures such as 'e-mail bombs', flooding the sites with large amounts of meaningless data. Access from China to the IP addresses of the group's Websites in the United States, Canada and the United Kingdom was also blocked.[42] In fact, the majority of charges for 'political' offences in cyberspace might well be in relation to Falungong. For example, several students of Beijing's Qinghua University were reportedly charged in 2000 and 2001 and sentenced to long jail terms for posting articles criticising government policy with respect to Falungong and for downloading and distributing material related to the movement.[43]

In general, this kind of censorship can be divided into proactive and reactive measures. The blocking of Websites is an obvious case of proactive censorship, and is applied mainly to Websites operated by foreign news services like CNN, the BBC or international human rights organisations. Judging from available reports, such blocking seems to be rather erratic and unsystematic. The *New York Times*, for example, was blocked until it published a lengthy interview with President Jiang Zemin.[44] It is sometimes possible to access the *International Herald Tribune*, the *Far Eastern Economic Review* and even *Human Rights in China*, but not always.[45] Blocks are sometimes temporarily lifted on special occasions. One such event was the October 2001 APEC summit in Shanghai, when foreign media sites were unblocked during a phase of (nearly) 'total digital freedom'.[46] It should also be noted that certain IP addresses have become unreachable at times due to bottlenecks in international data traffic. While bandwidth capacity has been enormously expanded during recent years, the number of Internet users in China has grown even faster.[47]

The phenomenon of Internet cafés also poses a problem for state control as they have become increasingly popular, especially in China's cities. During the sessions of the National People's Congress and the Chinese People's Political Consultative Conference in spring 2001, some delegates motioned for closing down all of these, arguing that they are harmful for children and young people. The discussion centred mainly on the 'online poisons' of pornographic material and online gambling.[48] Internet cafés have been raided several times by security forces, with crackdowns aimed at identifying places that are operating without a proper business licence and at stopping access to 'unhealthy' Websites, especially for the young.

In this respect, it should be noted that Internet cafés are not only required to apply for a business licence and register with the local Public Security

Bureau, but also to hire appropriate personnel to monitor the activities of users, who themselves are supposed to show an ID and register their details.[49] Following a decree from the State Council for a 'cleansing' campaign of Internet cafés between April and June 2001,[50] it was reported that some 8,014 institutions were closed, with about 2,000 of them permanently shut down and 6,000 temporarily undergoing 'rectification'. All in all, 56,800 Internet cafés were inspected on a national scale.[51] According to new rules issued in May 2001, Internet cafés are now required to be a minimum distance of 200 metres from government offices, army units and party organisations, as well as from primary and middle schools.[52]

Surveillance of the activities of ISPs has also begun to be more systematic. In March 2001, reports began to appear on sample checks conducted among four large ISPs in Shanghai to see if their customers held the necessary licences for their Internet presence. The result of the survey was declared to be satisfactory, with between 80 and 95 per cent of the Internet services registered as required.[53] The ability of the state to carry out such exercises is steadily being increased by beefing up the relevant personnel required to meet the challenges of the digital age. This includes the creation of special police units, which first appeared in Anhui province in August 2000, and have since spread to other regions and cities. These forces are charged with the central tasks of fighting cyber crime, ensuring IT security through work such as providing information and consultancy on computer viruses, and 'keeping order' in cyberspace.[54]

The physical ability of the state to surveil and punish is not necessarily the most important factor when it comes to maintaining 'security' in the realm of information technology, however. As Winkel points out, the concept of security in this field must be understood in terms of the 'objective security' that is derived from the reliability of social and technical functions, and the 'subjective security' that arises from the state of consciousness that is determined by individual perception and social communication. This becomes crucial in relation to the use of the Internet for political dissent, when we realise that 'The development of trust in a technology is a precondition for its acceptance, that is, for its being embraced and used by the people concerned'.[55]

The issue of trust may not be such a salient concern in liberal-democratic societies, where the monitoring activities of non-state actors are less likely to be associated with state intervention or coercive measures in the minds of users, and therefore ignite little protest. Most Internet users in the United States or Europe have probably never given much serious thought to the kinds of technological capacities for storing user data and carrying out surveillance that are available. They most likely do not care too much so long as there exists a significant degree of 'trust capital' in the operation of modern information technologies. Even if the individual user does wonder about what happens to the electronic traces he or she might leave in cyberspace, one would normally assume that these traces will not have any legal, let alone penal, consequences.

In an authoritarian state, where citizens have lived with censorship for decades, the issue might present itself in a different light. Doubts with respect to 'availability', but even more so with respect to the 'confidentiality' of electronic information, would suggest a lack of 'trust capital' and a good amount of insecurity concerning the capacities of the state (and its helpers) to watch what is going on in cyberspace. Just as in the Panopticon cited by Boyle and Foucault, the perception that one just might be under observation is likely to be more important in ensuring conformity with the rules of behaviour than whether one is actually being observed at any specific point in time. Such a vision could explain why the majority of censorship measures in Chinese cyberspace are in fact rather limited. Police forces and courts only need to become active sporadically if high-profile arrests and sentences can be made to constitute an effective deterrence by demonstrating the risks associated with dissident behaviour. The trick for the state is thus to nurture an attitude of 'voluntary' self-control and self-censorship among users, a 'firewall within one's head' as the *People's Daily* puts it.[56] The regulations need not necessarily be enforced in a strict sense to achieve this. In fact, it may actually be to the benefit of the state to leave a degree of vagueness in the terminology it uses, interpreting regulations in as loose or as strict a sense as is necessary to ensure that users and providers will always err on the side of caution when it comes to assessing the risks of dissent. For example, practically any kind of information can be declared to be a 'state secret' in China, even seemingly harmless statistical data on the last grain harvest.[57] In this way, laws become a mere supplement to much more subtle means of psychological control.

Role of the market

Another factor that plays a crucial role for exerting control over the Internet is the emerging alliance between the state and the official media, state-owned enterprises and financially strong investors. Qiu describes the emerging relationship well when he explains how China's Internet industry has become characterised by the rapid formation of an authoritarian-capitalist coalition that has seized the central spot that used to belong to small to medium-sized enterprises. The new arrivals are not only strengthened by traditional media conglomerates that serve as the mouthpiece of the party-state, but also by the close networking that takes place between wealthy investors and powerful political figures from both inside and outside the country that has developed in the context of commercialisation without political democratisation.[58]

The support of the state for the official media and large enterprises like Legend Computers in the IT sector, in addition to legal requirements that are imposed with respect to finances and personnel, leads to the marginalisation of smaller providers. Moreover, Internet firms that do not want to endanger their market position are well advised to play by the rules set by the state. Managers of Internet portals have thus expressed no surprise over the raft of regulations that were introduced in 2000, with one operator explaining the bottom line as being that, '[T]he new regulations don't make anyone happy, but they're completely expected and in line with other Chinese government policies. Anyone who was not already in de facto adherence with the policies was being very naïve'.[59] In fact, Internet providers had hurried to comply with draft versions of the regulations before they even came into force, revealing how they have to walk a tightrope between the interests of their customers and the demands of the state when operating in China. As an American representative of the Chinese-language portal Sina.com puts it, 'We are playing that role, to let people talk about sensitive issues but also to help the government manage the flow of ideas'.[60] Due to the vague wording of the regulations and their incoherent or arbitrary enforcement, Internet providers have to err on the side of caution when trying to decide what the authorities are prepared to tolerate.

This situation means that ICPs offering BBS services or chat rooms have to use the whole array of methods described by Boyle to indulge in self-censorship. Filter software that can block access to a certain Website or sound the alarm if it contains the word 'nipple' or one of its synonyms can also be programmed to do the same for the name of Taiwan's President or the word 'Falungong'. Such is the nature of 'neutral' technology, much of which is supplied by Western firms eager to supply the Chinese market with software and technology. Among others, Cisco Systems, Sun Microsystems, Nortel Networks (Canada), Dupont, and Daniel Data Systems (Israel) are known to have displayed their products at the trade fair 'Security China 2000', organised with the co-operation of the Ministry of Public Security. The advanced firewall software and other technical solutions offered by such firms is not only useful for protecting police networks from hacker attacks, but also for screening millions of electronic messages for key words. In fact, the focal point of the fair was a government project called 'Golden Shield'[61] that aims to safeguard the security of computer networks and fight cyber crime by connecting together the databases and surveillance systems of national and local police stations. It is ironic, therefore, that while the Western media frequently criticise China for obstructing the development of the Internet, it is Western firms that are supplying the technological means which enable China to carry out surveillance. In the meantime, Chinese companies have also started to develop special monitoring and filter software for themselves.[62] Early in 2001, for example, the Ministry of Public Security introduced an 'Internet Police 110' software package to be supplied in various versions for schools, individuals and Internet cafés that is supposed to filter out 'unhealthy' information, including not just violence and pornography but also anything positive about the Falungong or Tibetan exile and human rights groups.[63]

ISPs and ICPs employ special personnel, or 'big mamas', to look closely at the postings and contributions that filter software identifies as suspicious, in order to either delete them or clear them as being appropriate for the Internet. The personal decisions of Webmasters are thus a key element among the series of factors deciding on what gets onto the Internet and stays there.[64] One important form of 'punishment' for those who break the rules is simply to be eliminated from the Net, with no explanation necessary. Webmasters and system administrators can, moreover, criticise or reprimand authors for posting controversial contributions. If an author repeats his behaviour, then his or her IP address or registered name can be permanently blocked, amounting to the termination of their virtual existence.[65] While such punishments can be meted out in cyberspace, additional deterrence credibility is gained from the possibility of harsher 'real-world' sanctions in cases that are considered to be grave violations.

The Internet branches of the important official media probably enjoy more freedom than Internet portals – whose position is politically uncertain when it comes to deciding on the borders of what is allowed – precisely because their censors know what the government line is. This goes some way to explaining why a particularly popular forum for debate is the Strong State Forum (*Qiang guo luntan*) on the *People's Daily* Website.[66] While BBSs and the chat rooms of university networks or Internet portals seem to be closed down as a precaution before a sensitive anniversary such as 4 June, the date of the Tiananmen massacre, or when a new event triggers a fierce online debate, so far this has not happened to the Strong State Forum. An explosion at a school in Jiangxi province, for example, led to an intense discussion in cyberspace, with many postings calling into question the official government version of the incident, which blamed it on the actions of a lunatic. When too many comments with critical undertones appeared on Sina.com's online discussion forum, however, it was temporarily shut down.[67] Another method reported to have been used to stifle debate is to reduce data transfer speed before sensitive dates approach,[68] although this has the drawback of affecting state-run Websites as well.

Topics banned from public discussion include the Tiananmen Square events of June 1989, the Falungong movement and any explicit criticism of China's leaders. As regards current events such as scandals concerning official corruption, decisions are made on a case-by-case basis.[69] That there seems to be more toleration for comments on international incidents than on domestic affairs indicates that the Internet provides a welcome and officially tolerated outlet for nationalistic sentiments,[70] as long as the postings are not directed against the Chinese government and do not challenge state policies. The collision of the American EP-3 reconnaissance plane and a Chinese F-8 fighter off the Chinese coast in April 2001, for example, triggered a deluge of commentaries and personal statements in Chinese cyberspace. The overwhelming majority of these postings expressed a clear-cut anti-American attitude and most of them supported the position of the Chinese government. Within this context, one online comment pointed out that censorship measures were responsible for this nearly complete uniformity of opinions on the Net.[71] One analysis of postings made during this incident seems to corroborate the impression of uniformity, but also notes that a few critical contributions did appear, especially on the popular Chinese Internet portals Netease.com and Sina.com.[72]

As was to be expected, the terrorist attacks of 11 September 2001 in the United States sparked a hot debate in China as well. In the days directly following the attacks, the majority of the postings on Strong State Forum, despite deploring the death of so many people, expressed the view that these attacks were the logical result of the unilateral and 'hegemonistic' policy of the United States and that terrorism was nothing but the reverse side of hegemony and power politics. Others argued that the terrorist attacks were the outcome of a type of globalisation that, under the leadership of the United States, had widened the gap between rich and poor in the world.[73] That appeals for people to refrain from downright anti-American postings began to appear after a few days should be understood in the context of the Chinese leadership's decision to interpret the formation of the United States-led anti-terrorist coalition as an opportunity to improve Sino-American relations, making a wave of anti-American sentiment within the population less desirable than before.

Counter strategies

It would be unrealistic to claim that the array of methods used by the state to ensure its control over the Internet provide a watertight system, when spaces for the development of counter strategies certainly do exist. For example, it is common in China for regulations to be only loosely enforced. Some Internet cafés also operate without the required business licences, making them less likely to insist on the proper identification and registration of their customers. Moreover, Internet users can and do find ways to outsmart government blockades and restrictions through methods such as using proxy-servers outside China to access banned addresses.[74] Lists of the IP addresses of proxy-servers are reportedly distributed among Internet users in China, sometimes via e-mail. Members of the Falungong movement are also supposed to have learned how to circumvent blockades and protect their electronic communication using encryption programmes.[75] There has been at least one occasion on which commercial interests appear to have weakened the hold of the state, when the government tried to gain control over the usage of encryption by passing a new regulation in December 1999. Strong objections from Western firms like Microsoft led to the measure being considerably watered down.[76]

However, the state has so far also proved resourceful in its attempts to close such gaps. The IP addresses of proxy-servers can themselves be blocked, for example. Security agencies are even said to circulate false IP-addresses for such servers, so that attempts to get round blockades might end up being routed directly to the Public Security Bureau.[77] Moreover, the international priority given to strengthening surveillance of the Internet following 11 September does not seem to be encouraging such challenges to state control by foreign commercial actors. This point is well illustrated by the case of

SafeWeb, a United States-based software company that developed 'Triangle Boy', a method for maintaining anonymity in cyberspace. In August 2001, it was reported that International Broadcasting Bureau, the parent company of Voice of America, had entered into negotiations with SafeWeb to finance a project to undermine China's efforts to censor the Internet.[78] Soon after 11 September, this venture ran into problems as the atmosphere in the United States became far more sympathetic to efforts to strengthen online surveillance, although SafeWeb is reported to be keeping its services going for Voice of America on a trial basis.[79]

An additional advantage for the state is that a certain level of technical knowledge and sophistication is necessary for employing the methods of circumvention described above, and even users with the requisite expertise have to ask themselves whether they are willing to take the risk involved in actually using such skills. According to official statistics, the vast majority of Internet users are young male urban citizens with more than average education and more than average income.[80] Their main motives for accessing the Internet are to gather information, access educational services and for entertainment. Although surveys on Internet usage in China conducted since 1997 used to reveal that lack of Chinese-language information on the Internet was one of the most frequently mentioned complaints, this problem has now gone down in importance, not least due to state-sponsored initiatives such as 'Government Online' and the development of a significant Web presence for the 'traditional' state media. This expansion and improvement of information and entertainment offered in the Chinese language, largely provided on the Internet by organisations with strong links to the state, clearly reduces the need to look for alternative sources and activities in cyberspace.[81]

It remains far from clear just how interested the average Chinese Internet user is in accessing information and activities that could be deemed politically subversive, with empirical evidence still very sketchy. At least two surveys have been conducted on opinions in urban areas,[82] but many more in-depth analyses will be necessary to get a clearer picture of online behaviour in China and of how the different forms of electronic communication are used and how this relates to the off-line world. Statements made by young people and students suggest that they see accessing blocked sites as a kind of game, with politics not being their core interest.[83] A strong enough motivation seriously to risk not only one's 'virtual' existence but also one's real existence by engaging in dissenting activities on the Internet can only be readily assumed for groups like the Falungong, which operate outside the law anyway. There is thus a real danger that the impression generated by the Western media of Chinese cyberspace as the stage for a battle between a repressive state and Internet users ceaselessly posting and hunting for politically subversive information restricts our perspective to a very small and possibly insignificant aspect of the overall situation. The majority of observers who have followed and analysed the development of the Internet in China in more depth do not support the picture of a state rendered powerless over an uncontrollable

Internet, but tend to conclude that the authorities are able to exert their control over online users as much by simple intimidation as by sophisticated electronic surveillance or by blocking direct access to politically suspect foreign Websites.[84]

Is resistance futile?

In light of the above evidence, it is safest to conclude that while the Internet in China cannot be protected from every form of subversive usage by insurmountable 'digital fences', state control and censorship have not simply evaporated either. Instead, the state is meeting the challenges of the digital age by combining the kinds of factors of control elaborated on by writers like Lessig and Boyle. This involves a complex interplay between the state and the key commercial actors in the sector, namely ISPs, ICPs and the official media, most of which are partly or wholly owned by the state anyway. By practising self-censorship and executing the duties of surveillance and supervision assigned to them by the state, such actors partially relieve it of the task of control and censorship. All this becomes possible through the technological embedding of control in the architecture of the Internet itself, which is nothing specific to China, but mainly the result of the commercialisation of the Internet worldwide. The introduction of regulations is only really necessary to complement this strategy by increasing the deterrence effect of feeling that one's actions just might be under observation and by making examples of high-profile cases in the courts.

This does not mean that the Internet is politically irrelevant, just that it is not likely to be the cause of significant social change. Instead, freedom of expression in China has expanded considerably since the end of the late 1970s due to the overall policy of 'reform and opening' initiated by Deng Xiaoping, despite intermittent phases of contraction.[85] Thus, what can be said of the Internet can be said of other media in China as well, namely that the limits of toleration are constantly being tested and re-negotiated. The common practice of conducting activities under the cloak of pseudonyms in Chinese cyberspace and the relative weakness of 'virtual' sanctions might still make the Internet more of a catalyst of social change than other media, but it is most likely to play a significant role if a social or political movement emerges in the non-virtual world. It is then that, along with fax machines, mobile phones, and mails via mobile phones, the Internet's ability to distribute news and facilitate organisation could play a decisive role. However, there is little reason to believe in the kind of technological determinism that postulates that the Internet could trigger a democratic or other mass movement in China by itself. The Internet might ignore territorial boundaries or surmount them without much effort (although even this has begun to change), but this does not mean that it exists in a social and political vacuum, detached and independent of its environment.[86]

Notes

1 Clinton's statement is cited in W.J. Drake, S. Kalathil and T.C. Boas, 'Dictatorships in the Digital Age: Some Considerations on the Internet in China and Cuba', *iMP*, October 2000. Online. Available HTTP: <http://www.cisp.org/imp/october_2000/10_00drake.htm> (accessed 11 November 2000).
2 See 'Jiang Looks to Information Technology to Drive Economy', *China IT and Telecom Report*, 25 August 2000, vol.1, no.45, pp. 4–5; A. Lin Neumann, 'The Great Firewall', *CPJ Briefings: Press Freedom Reports*. Online. Available HTTP: <http://www.cpj.org/Briefings/2001/China_jan01/China_jan01.html> (accessed 23 February 2001). In March 2001, an article in *People's Daily* divided 'unhealthy' Internet contents into 'black' (false information from hostile forces, intended to 'destroy' and 'westernise' China), 'grey' (a flood of meaningless blabbing, negative sentiments and vulgar thoughts) and 'yellow pollution' (pornography). See 'Party Daily Calls for Action Against Internet Pollution', *Renmin Ribao* Website, 21 March 2001, in *BBC Summary of World Broadcasts – Far East* (*SWB FE*), 28 March 2001, no. 4106, p. G/7.
3 See Chapter 1 by Xiudian Dai in this book for more details on this issue.
4 Dali L. Yang, 'The Great Net of China'. Online. Available HTTP: <http://www.mfcinsight.com/article/010209/oped4.html> (accessed 16 February 2001); Zhang Junhua: 'China's "Government Online" and Attempts to Gain Technical Legitimacy', *Asien*, July 2001, no. 80, pp. 93–115.
5 '*Wang ju ren de liliang*', literally 'the Net concentrates the power of people'. See Chinese Website of Netease. Online. Available HTTP: http://news.163.com/editor/010220/010220_109937.html (accessed 26 March 2001).
6 For a summary of the arguments of 'digital libertarianism', see J. Boyle, 'Foucault in Cyberspace: Surveillance, Sovereignty, and Hard-Wired Censors', 1997. Online. Available HTTP: <http://www.wcl.american.edu/pub/faculty/boyle/foucault.htm> (accessed 6 November 2000). A definition of 'digital libertarianism' can be found under http://www.feedmag.com/html/feedline/97.09marshall/97.09marshall1.2.html (accessed 19 April 2001): the proponents of this approach argue, on the one hand, in favour of the non-interference of the state in cyberspace, while on the other they claim that the Internet is characterised by certain features which are beyond the control of the state. Three main groups can be identified: computer enthusiasts (hackers who want to keep 'their' Internet for themselves); traditional liberals who want to limit state interference; and economically oriented conservatives, who consider the global reach of the Internet as desirable since it weakens the regulatory powers of governments.
7 See, for example, N. Robinson, 'New Laws Seek to Balance Privacy and Surveillance', *Jane's Intelligence Review*, January 2002, vol. 14, no. 1, pp. 52–3; C.S. Kaplan, 'Concern Over Proposed Changes in Internet Surveillance', 21 September 2002. Online. Available HTTP: <http://www.nytimes.com/2001/09/21/technology/21CYBERLAW.html> (accessed 21 September 2001).
8 See, for example, 'Putting It In Its Place' and 'The Internet's New Borders', *The Economist*, 11 August 2001, pp. 18–20 and pp. 9–10; A.E. Cha, 'Bye-Bye Borderless Web: Countries Are Raising Electronic Fences', *International Herald Tribune*, 5 January 2002, pp. 1, 4.
9 'Putting It In Its Place', p. 18.
10 L. Lessig, *Code and Other Laws of Cyberspace*, New York: Basic Books, 1999.
11 Ibid., p. 89.
12 Ibid., p. 57.
13 Ibid., p. 60.
14 Ibid., p. 97.

15 For the following observations see Boyle, 'Foucault in Cyberspace'.
16 M. Foucault, *Discipline and Punish. The Birth of the Prison*, London: Penguin, 1991. Especially Part 3, Chapter 3: 'Panopticism', pp. 195–228.
17 Boyle, 'Foucault in Cyberspace'.
18 For United States initiatives, such as the Clipper Chip, see Lessig, *Code and Other Laws of Cyberspace*, pp. 47ff.; A.L. Shapiro, 'The Internet', *Foreign Policy*, Summer 1999, no. 115, pp. 18–19. On a controversial draft in England in 2000 see C. Grande, 'Uunet and Nokia Attack E-mail Legislation', *Financial Times*, 12 July 2000, p. 12. Providing the state with access to encryption has been compared to forcing all citizens to deposit a key to their house at the police station, so that security forces could enter it if they suspect a burglary has taken place.
19 J. Linchuan Qiu, 'Virtual Censorship in China: Keeping the Gate between the Cyberspaces', *International Journal of Communications Law and Policy*, Winter 1999/2000, issue 4. Online. Available HTTP: <http://111.ijclp.org/4_2000/ijclp_webdoc_1_4_2000.html> (accessed 10 November 2000), p. 22.
20 Ministry of Information Industry, '*Zhonghua Renmin Gongheguo dianxin tiaoli*' (Telecommunication Regulations of the People's Republic of China), 25 September 2000, *Guowuyuan gongbao* (*GWYGB*), 2000, no. 33, pp. 11–21. Also Online. Available HTTP: <http://www.mii.gov.cn/news2000/1013_1.htm> (accessed 2 December 2000); State Council, '*Hulianwang xinxi fuwu guanli banfa*' ('Methods for the Administration of Internet-Based Information Services'), 20 September 2000, *GWYGB*, 2000, no. 34, pp. 7–9. Also in *People's Daily*, 7 November 2000. Online. Available HTTP: <http://www.peopledaily.com.cn/GB/channel5/28/200010017/2557566.html> (accessed 24 October 2000); State Council Information Office, Ministry of Information Industry, '*Hulianwangzhan congshi dengzai xinwen yewu guanli zhanxing guiding*' ('Interim Provisions for the Administration of Release of News by Websites'), 6 November 2000, GWYGB, 2001, no. 2, pp. 46–8; Ministry of Information Industry, '*Hulianwang dianzi gonggao fuwu guanli guiding*' ('Provisions for the Administration of Electronic Information Services on the Internet'), 8 October 2000, *GWYGB*, 2001, no. 2, pp. 45–6; National People's Congress Standing Committee, '*Guanyu weihu hulianwang anquan de jueding*' ('Resolution of the Standing Committee of the National People's Congress on Maintaining Security of Computer Networks'), 28 December 2000, *GWYGB*, 2001, no. 5, pp. 21–3. An excellent summary and evaluation of the main contents can be found in K. Giese, 'Das gesetzliche Korsett für das Internet ist eng geschnürt', *China aktuell*, October 2000, pp. 1173–81.
21 On the administrative structure see Chapter 1 by Xiudian Dai in this volume. In addition, a whole number of ministries, government institutions and administrative units – such as the Ministry of Public Security, the ministries of Culture and of Health, the State Administration for Industry and Commerce and many more – have a say in drafting Internet regulations relevant for their respective domains.
22 This list can be found as § 15 in '*Hulianwang xinxi fuwu guanli banfa*', § 9 in '*Hulianwang dianzi gonggao fuwu guanli guiding*' and § 13 in '*Hulianwangzhan congshi dengzai xinwen yewu guanli zhanxing guiding*'.
23 State Council, '*Chuban guanli tiaoli*' ('Regulations Governing the Administration of the Publishing Industry'), 1 January 1997, *GWYGB*, 1997, no. 2, pp. 38–46. In fact, in these regulations electronic publications were already explicitly mentioned as falling under their jurisdiction (§ 2). The banned contents, which also include 'superstition', can be found in § 25. New regulations for the publishing industry were issued in December 2001. They contain the complete list of forbidden contents cited above; State Council, '*Chuban guanli tiaoli*'

('Regulations Governing the Administration of the Publishing Industry'), 25 December 2001, *GWYGB*, 2002, no. 4, pp. 14–20.
24 G. Chen and D.A. Britton, 'China To Allow ICPs To Report Their Own News', 24 April 2000. Online. Available HTTP: <http://www.chinaonline.com/topstories/000424/1/C00042420.asp> (accessed 25 April 2000); 'China: Eastsay.com May Not Write, Collect News', 5 June 2000. Online. Available HTTP: <http://www.chinaonline.com/topstories/000605/1/c00060203.asp> (accessed 6 June 2000).
25 'China's Print Media Concerned Over New Internet Portals', 22 September 1999. Online. Available HTTP: <http://www.chinaonline.com/issues/legal/currentnews/open/C9092180e-SS.asp> (accessed 24 October 1999); G. Chen, 'China's Booming Internet Sector: Open or Closed To Foreign Investment?', 8 October 1999. Online. Available HTTP: <http://www.chinaonline.com/industry/infotech/NewsArchive/Secure/1999/october/C9100519REV-SS.asp> (accessed 9 January 2000). In Shanghai, the most important traditional media (*Jiefang Ribao*, *Wenhui bao*, *Xinmin Wanbao*, Shanghai TV Station and others) united to form their own Internet portal. See 'China Web Site Has Got It Covered – Newswise', 24 May 2000. Online. Available HTTP: <http://www.chinaonline.com/issues/internet_policy/NewsArchive/Secure/2000/may/b100052237.asp> (accessed 19 April 2001).
26 T. Fravel, 'The Bureaucrats' Battle over the Internet in China', 17 February 2000, Online. Available HTTP: <http://www.virtualchina.com/news/feb00/021800-ministries-tf.html> (accessed 19 February 2000). This is especially noteworthy if seen against the background that the CCP has drastically reduced the subsidies for newspaper publishing companies during the past years and that the publishing companies are strongly encouraged to finance themselves. See in detail D. Fischer, 'Rückzug des Staates aus dem chinesischen Mediensektor? Neue institutionelle Arrangements am Beispiel des Zeitungsmarktes', paper presented at the first meeting of ASC (Arbeitskreis sozialwissenschaftliche Chinaforschung) 'Funktionswandel des Staates', 17–18 November 2000 in Witten.
27 A portrait of this institution can be found under: 'Internet Information Management Bureau (IIMB)', 30 May 2000 Online. Available HTTP: <http://www.chinaonline.com/refer/ministry_profiles/IIMB.asp> (accessed 2 November 2000).
28 Regulations for electronic information services can be found in §§ 14 and 15 of '*Hulianwang dianzi gonggao fuwu guanli guiding*', p.46.
29 See, for example, V. Pik-Kwan Chan, 'Beijing in Hard Drive to Patrol Net Users', 18 January 2002, online, available HTTP: <http://china.scmp.com/ZZZTHW3M5WC.html> (accessed 18 January 2002).
30 '*Hulianwang dianzi gonggao fuwu guanli guiding*', p.46 (§ 10).
31 Ibid.
32 This difference is stressed by Qiu, 'Virtual Censorship in China'. On Singapore's Internet policy in detail G. Rodan, 'The Internet and Political Control in Singapore', *Political Science Quarterly*, 1998, vol. 113, no. 1, pp. 63–89.
33 References to §§ 21 and 22 in '*Hulianwang xinxi fuwu guanli banfa*'.
34 §§ 22 and 23 of '*Hulianwang xinxi fuwu guanli banfa*'.
35 According to the penal law of the PRC, sentences for several years are possible. See Giese, 'Das gesetzliche Korsett für das Internet ist eng geschnürt', p. 1,176.
36 See also Giese, 'Das gesetzliche Korsett für das Internet ist eng geschnürt', pp. 1,176ff.
37 The number of cases varies depending on the source used. J. Linchuan Qiu, 'Internet Censorship in China (1999–2000)', *Communications Law in Transition Newsletter*, 18 February 2001, vol. 2, no. 3. Online. Available HTTP: <http://pcmlp.socleg.ox.ac.uk/transition/issue2_3/qiu.htm> (accessed 25

February 2002). Qiu states that due to the increase in specialised police units and the deployment of new technologies, 13 people were arrested between summer 1999 and the end of 2000, while only one person was arrested in the years prior to that date. According to Neumann, 'The Great Firewall', 7 people were arrested for Internet 'crimes' between 1998 and 2000. A *Human Rights Watch Backgrounder* published in July 2001 lists 15 individuals detained for posting material on the Internet. See 'Freedom of Expression and the Internet in China'. Online. Available HTTP: <http://www.hrw.org/backgrounder/asia/china-bck-0701.pdf> (accessed 24 August 2001).
38 'The Cracker War on China', 15 January 1999. Online. Available HTTP: <http://www.virtualchina.com/infotech/perspectives/perspective-011599.html> (accessed 2 January 2000); 'Whither the China Net?', 5 February 1999. Online. available HTTP: <http://www.virtualchina.com/infotech/perspectives/perspective-020599.html> (accessed 2 January 2000); 'Chinese Engineer who Helped Dissident Newsletter is Freed', *Digital Freedom Network*, 6 March 2000. Online. Available HTTP: <http://www.dfn.org/focus/china/linhai.htm> (accessed 28 February 2001).
39 'Trial of Internet Entrepreneur Starts in Chengdu', RTHK Radio 3 audio Website, Hong Kong, 13 February 2001, cf. *SWB FE*, 14 February 2001, no. 4070, p. G/4. See on this case also 'Chinese Webmaster to be Tried February 13', 9 February 2001, *Digital Freedom Network*. Online. Available HTTP: <http://www.dfn.org/focus/china/huangqi-trialdate.htm> (accessed 28 February 2001); '"6-4" Web Site Creator Put to Subversion Trial in Sichuan', 11 February 2001, *China News Digest*, no. GL01-019, available e-mail: LISTSERV@LISTSERV.ACSU.BUFFALO.EDU (12 February 2001).
40 'Chinese Internet Café Owner Arrested', 24 August 2000, *Digital Freedom Network*. Online. Available HTTP: <http://www.dfn.org/focus/china/jiangshihua.htm> (accessed 28 February 2001).
41 'Banned Chinese Web site Reappears', 11 August 2000, Digital Freedom Network. Online. Available HTTP: <http://www.dfn.org/focus/china/xinwenmingback.htm> (accessed 28 February 2001).
42 C.S. Smith, 'Falun Dafa Defies Authority by Preaching in Cyberspace', *Asian Wall Street Journal*, 10–11 September 1999, pp. 1, 6; I. Buruma, 'China in Cyberspace', *New York Review of Books*, 4 November 1999, vol. 46, no. 17, pp. 9–12; 'Falun Dafa and the Internet: A Marriage Made in Web Heaven', 30 July 1999. Online. Available HTTP: <http://www.virtualchina.com/infotech/perspectives/perspective-073099.html> (accessed 2 January 2000); 'Government, Falun Gong followers in Internet battle', *Sing Tao Jih Pao* (Hong Kong), 28 July 1999, p.A4, cf. *SWB FE*, 29 July 1999, no. 3599, pp. G/5–6.
43 'Sentencing Postponed in "Cult" Web Case', 19 Februray 2002. Online. Available HTTP: <http://uk.news.yahoo.com/020219/80/csj9z.html> (accessed 22 February 2002).
44 J. Lee, 'U.S. May Help Chinese Evade Net Censorship', *New York Times*, 30 August 2001. Online. Available HTTP: <http://www.nytimes.com/2001/08/30/technology/30VOIC.html> (accessed 31 August 2001).
45 A. Lin Neumann, 'The Great Firewall'.
46 'China Eases Net Censorship During APEC Talks', *South China Morning Post*, 17 October 2001. Online. Available HTTP: <http://technology.scmp.com/internet/ZZZMEIMFRSC.html> (accessed 17 October 2001).
47 Slow speed of Internet access and data transfer is the most frequent complaint of Internet users in China. International bandwidth was expanded by 163 per cent between July 1998 and December 1999 (from 84.64 to 351 Mbps). Since the number of Internet users grew by 291 per cent during the same time, there was no improvement of the situation. However, during the year 2000, this trend was

reversed: while international bandwidth grew again by 370 per cent, the number of users increased by 'only' 123 per cent. (Calculated on the basis of statistical data provided by CNNIC, the reliability of which is questioned by Giese in Chapter 2 in this volume). See also Qiu, 'Virtual Censorship in China', p. 7.
48 'Should Internet Cafes Be Closed?', *Beijing Review*, 26 April 2001, pp. 30–1.
49 'China Issues New Regulations for Internet Cafes', 21 January 1999. Online. Available HTTP: <http://www.chinaonline.com/industry/infotech/NewsArchive/Secure/1998/August/it_b8081003e.asp> (accessed 27 January 2000); 'China's Beijing Cracks Down on Internet Cafes', 24 March 2000. Online. Available HTTP: <http://www.chinaonline.com/topstories/000324/2B200032207.asp> (accessed 29 March 2000).
50 'State Council Tightens Control over Internet Cafes', 17 April 2001. Online. Available HTTP: <http://www.chinaonline.com/topstories/010417/1/C0104201.asp> (accessed 18 April 2001). On the crackdowns in 2001: 'Beijing Police Pull Plug on Illegal Internet Cafés', 23 February 2001. Online. Available HTTP: <http://www.chinaonline.com/topstories/010223/1/C01021611.asp> (accessed 26 February 2001). According to this report, in the year 2000, in Beijing's Chaoyang district alone, 41 illegal Internet cafés were closed and had to pay RMB 350,000 fines, and 11 Internet cafés were reprimanded and temporarily closed. In the first two months of 2001, the police found 5 more illegal businesses. See also 'Hubei's Daye City Closes Cybercafés for Pornography, Falun Gong Information', Hubei Radio Website, 10 February 2001, cf. *BBC SWB FE*, 16 February 2001, no. 4072, pp. G/7–8.
51 'China Shuts Down Nearly 2,000 Internet Cafes', 19 July 2001. Online. Available HTTP: <http://asia.dailynews.yahoo.com/headlines/regional/china.html> (accessed 19 July 2001).
52 Michael Ma, 'Beijing gets Tough on Internet Bar and BBS Operators', *South China Morning Post*, 10 May 2001. Online. Available HTTP: <http://china.scmp.com/today/ZZZY0FMKYLC.html> (accessed 10 May 2001).
53 'Shanghai Conducts Random Check of Web Site Operators', 20 March 2001. Online. Available HTTP: <http://www.chinaonline.com/topstories/010320/1/b101030934.asp> (accessed 21 March 2001).
54 See Neumann, 'The Great Firewall'; 'Internet Police Ranks Swell to 300,000', *Ming Pao* Website, 8 December 2000. See also *BBC SWB FE*, 11 December 2000, no. 4020, pp. G/5–7.
55 O. Winkel, 'Sicherheit in der digitalen Informationsgesellschaft', *Aus Politik und Zeitgeschichte*, 6 October 2000, no. B 41–42, p. 21.
56 The idea of a 'firewall' within one own's head was explicitly advocated in an article on the Website of *Renmin Ribao*. 'Party Daily Calls for Action Against Internet Pollution', 21 March 2001, in *BBC SWB FE*, 28 March 2001, no. 4106, p. G/7.
57 'Bureau for the Protection of State Secrets (State Secrets Bureau)', 28 January 2000. Online. Available HTTP: <http://www.chinaonline.com/refer/ministry_profiles/Secrets-3-S.asp> (accessed 31 January 2000). The State Secrets Law came into force in 1988. 'Provisions Governing the Implementation of the State Secrets Law of the People's Republic of China' was issued in 1990. State Secrets Bureau, '*Zhonghua Renmin Gongheguo baoshou guojia mimi fa shishi banfa*', *GWYGB*, 1990, no.14, pp. 538–43. The regulations define under what circumstances any information is considered to be a state secret. One of the points listed here refers to information which 'undermines consolidation and defence of the State's political power and which influences the unity of the State, ethnic unity and social stability' (p. 538).
58 Qiu, 'Internet Censorship in China (1999–2000)'. Commercialisation changed the nature of the Internet in ways that mean that earlier claims can no longer be

unconditionally sustained. Cyberspace now allows practically everyone to become both a recipient *and* a provider of information, since no costly means of production are needed (in contrast to traditional media) and market entry costs are extremely low.
59 Commentary by representative of an Internet portal, cited in Neumann, 'The Great Firewall'. See also D. Cowhig, 'New Net Rules Not a Nuisance?', 5 December 2000. Online. Available HTTP: <http://www.chinaonline.com/commentary_analysis/internet/NewsArchive/secure/2000/December/c00120160.asp> (accessed 6 December 2000).
60 Cited in D.C. McGill, 'Sina.com's Delicate Balancing Act', 23 May 2000. Online. Available HTTP: <http://www.virtualchina.com/finance/stirfry/052300-stirfry-dcm-alo2.html> (accessed 25 May 2000).
61 On the 'Golden Shield' project see G. Walton, 'China's Golden Shield: Corporations and the Development of Surveillance Technology in the People's Republic of China', Rights and Democracy We site. Online. Available HTTP: <http://www.ichrdd.ca/111/english/contentsEnglish.html> (accessed 21 October 2001). Western companies also compete to help Saudi Arabia in blocking the Internet. See J. Lee, 'Companies Compete to Provide Saudi Internet Veil', *New York Times*, 19 November 2001. Online. Available HTTP: <http://www.nytimes.com/2001/11/19/technology/19SAUD.html> (accessed 19 November 2001). The first 'golden' projects were launched in 1994 in order to 'informatise' China – such as the 'Golden Card', 'Golden Customs', 'Golden Tax' and 'Golden Sea'. See N. Hachigian, 'China and the Net: A Love-Hate Relationship, Part II'. Online. Available HTTP: <http://www.chinaonline.com/commentary/archive/secure/2011/March/c01030260.asp> (accessed 8 March 2001); M. Mueller and Zixiang Tan, *China in the Information Age: Telecommunications and the Dilemmas of Reform*, Westport, CT and London: Praeger Publishers, 1997, pp. 45–64.
62 On the implications for national security arising from Chinese dependency of foreign software and hardware see Chapter 7 by Christopher R. Hughes in this volume.
63 'Spy Systems on Show in China', *Far Eastern Economic Review*, 2 November 2000, p. 10; '"Purifying" the Net, China-style'. Online. Available HTTP: <http://www.dfn.org/focus/china/filteringsw.htm> (accessed 28 February 2001); D. Gebler, 'Chinese Web Filter May Block Western Sites'. Online. Available HTTP: <http://www.newsfactor.com/perl/printer/7805/> (accessed 28 February 2001); M. Fackler, 'The Great Fire Wall of China?', 8 November 2000. Online. Available HTTP: <http://www.abcnews.go.com/sections/tech/DailyNews/chinanet001108.html> (accessed 28 February 2001); A.C. LoBaido, 'Life with Beijing's Bruisers'. Online. Available HTTP: <http://www.worldnetdaily.com/news/printer-friendly.asp?ARTICLE_ID=21446> (accessed 28 February 2001). On the question of importing Western hardware and software for surveillance see also F. Sieren, 'Von Netzen und Mauern. Über die Substanz chinesischer Internetphantasien', in K. Leggewie and C. Maar (eds), *Internet & Politik. Von der Zuschauer- zur Beteiligungsdemokratie*, Cologne: Bollmann, 1998, pp. 229–35.
64 A representative of Sohu.com described the approach as follows: 'We go with our intuition ... If something makes us uncomfortable, we nix it.' Cited in T. Marshall and A. Kuhn, 'China Goes One-on-One With the Net', *LA Times*, 27 January 2001. Online. Available HTTP: <http://www.latimes.com/business/cutting/lat_chitek010127.htm> (accessed 29 January 2001).
65 For a detailed description of this method see Wenzhao Tao, 'Censorship and Protest: The Regulation of BBS in China People Daily', *first monday*. Online.

Available HTTP: <http://www.firstmonday.dk/issues/issue6_1/tao/index.html> (accessed 23 February 2001); Qiu, 'Virtual Censorship in China'.
66 To be found on the *People's Daily* homepage, http://www.peopledaily.com.cn. The rules governing postings to *Qiang guo* are published on this site as well. No postings are permitted, for example, that 'contradict reform and opening as well as the four basic principles'. The same is true for unconfirmed news. Accounts of personal experiences are to be marked as such. No names of leading members of Party and government or other prominent personalities are to be used as pen names (*biming*). See <http://www.qglt.com/wsrmlt/rule.html> (accessed 7 March 2001).
67 'Top Website Shuts Chatroom over School Blast Anger', 9 March 2001. Online. Available HTTP: <http://www.ptdprolog.net/Webnews/wed/bo/Qchina-blast-internet.Rp0__BM9.html> (accessed 19 April 2001). It remained unclear, however, whether this closure was ordered by the authorities or was based on an in-house decision.
68 'China Fetes Falun Gong Day With Slow Bandwidth', 15 May 2000. Online. Available HTTP: <http://www.chinaonline.com/topstories/000515/x/C0051520.asp> (accessed 16 May 2000).
69 Wenzhao Tao, 'Censorship and Protest'.
70 On the question of Chinese nationalism on the Internet see C.R. Hughes, 'Nationalism in Chinese Cyberspace', *Cambridge Review of International Affairs*, Spring/Summer 2000, vol. 13, no. 2, pp. 195–209; A.R. Kluver, 'New Media and the End of Nationalism: China and the US in a War of Words', *Mots Pluriels*, August 2001, no. 18. Online. Available HTTP: <http://www.arts.uwa.edu.au/MotsPluriels/MP1801ak.html> (accessed 29 August 2001).
71 English translation of the Chinese posting: 'PRC Chatroom Censorship Criticized – Chatroom discussion', H-ASIA, available as e-mail: H-ASIA@H-NET.MSU.EDU (11 April 2001). The author makes the criticism that contributions which criticised the reaction of the Chinese government to the incident for being too lenient were removed from the Net. This critical message itself was able to survive on the Net for a while, however.
72 See J. Linchuan Qiu, 'Chinese Opinions Collide Online', *Online Journalism Review*. Online. Available HTTP: <http://ojr.usc.edu/content/story.cfm?request=561> (accessed 17 April 2001).
73 G. Wacker, 'Chinesische Reaktionen auf die Terroranschläge in den USA'. Online. Available HTTP: <http://www.swp-berlin.org/produkte/brennpunkte/wnd11sep6C.htm> (accessed 21 November 2001); N.D. Kristof, 'The Chip on China's Shoulder', *New York Times*, 18 January 2002. Online. Available HTTP: <http://www.nytimes.com/2002/01/18/opinion/18KRIS.html> (accessed 18 January 2002).
74 This involves the user maintaining anonymity by accessing the site in question by using a proxy-server to make a connection.
75 I. Johnson, 'Falun Gong Faces Added Pressure As Crackdown Grows', *Asian Wall Street Journal*, 28 March 2001, pp. 1, 6; C.S. Smith, 'Sect Clings to the Web in the Face of Beijing's Ban', *New York Times*, 5 July 2001. Online. Available HTTP: <http://www.nytimes.com/2001/07/05/world/05FALU.html> (accessed 7 July 2001).
76 State Council, '*Shangyong mima guanli tiaoli*' ('Regulations for the Administration of Commercial Encryption'), Directive No. 273, December 1999, *GWYGB*, 1999, no. 36, pp. 1663–7. The regulations, which came into force on 31 January 2000, stipulate that all companies must register encryption products and any equipment that uses such products. As a result of the protests, the regulations, which had been rather unclear to begin with, were revised in the

sense that they would not apply to mobile phones, Windows software and Internet browsers. See K. Marlow, 'China Softens Encryption Rules', 14 March 2000. Online. Available HTTP: <http://www.chinaonline.com/topstories/000314/1/C00031430.asp> (accessed 15 March 2000). On the question of encryption in general see Lessig, *Code and Other Laws of Cyberspace*, pp. 35ff.
77 This is mentioned in Qiu, 'Internet Censorship in China (1999–2000)'; J.M. Chen, 'Willing Partners to Repression?', 27 November 2000, *Digital Freedom Network*. Online. Available HTTP: <http://www.dfn.org/focus/china/multinationals.htm> (accessed 28 February 2001).
78 J. Lee, 'U.S. May Help Chinese Evade Net Censorship', *New York Times*, 30 August 2001. Online. Available HTTP: <http://www.nytimes.com/2001/08/30/technology/30VOIC.html> (accessed 30 August 2001).
79 E. Mills Abreu, 'SafeWeb Shuts Free Web Service Catering to Users in China', Chinese Internet Research, 20 November 2001. Online. E-mail posting. Available at: chineseinternetresearch@egroups.com (21 November 2001).
80 On the details of Internet distribution in China see Chapter 2 by Karsten Giese in this volume.
81 The Website 'FM365.com', which is sponsored by the biggest computer company in China, Legend Computers, offered an online game just in time for the National People's Congress in March 2001. The player could choose one of four officials accused of corruption (Yang Xianqian, Cheng Kejie, Wang Baosen, and Hu Changqing) and shoot at them by mouse-click. Moreover, the site offered background information on the four cases and even the option to give a vote on questions such as 'How do you judge the government's efforts to fight top-level corruption?' Possible answers were 'not enough by far', 'so-so' (*mama-huhu*), 'quite ok' and 'cannot say'. Online. Available HTTP: <http://www.fm356.com/shequ/> (accessed March 28 2001). In real life, three of the officials were sentenced to death and the fourth (Wang Baosen) committed suicide. The question whether this game could have gone online without official approval was briefly discussed in March 2001 in the mailing list 'Chinese Internet Research'. From a Western perspective, such a game might seem macabre, especially because it featured real individuals. It is remarkable, however, how the game 'Corrupt Officials' combined entertainment, information and popular feedback.
82 Guo Liang and Bu Wei, '*2000 nian Beijing, Shanghai, Guangzhou, Chengdu, Changsha Qingshao nian hulianwang shiyong zhuangkuang ji yingxiang de tiaocha baogao*' (Survey Report on Internet Usage and Influence in Beijing, Shanghai, Guangzhou, Chengdu and Changsha in the year 2000), April 2001. Online. Available HTTP: <http://www.chinace.org/ce/itre/index_.htm> (accessed 9 August 2001); J.J.H. Zhu and Zhou He, 'Information Accessibility, User Sophistication, and Source Credibility: The Impact of the Internet on Value Orientations in Mainland China', *Journal of Computer-Mediated Communication*, January 2002, vol. 7, no. 2. Online. Available HTTP: <http://www.ascusc.org/jcmc/vol/issue2/china.html> (accessed 21 February 2002).
83 Neumann, 'The Great Firewall'.
84 Marshall and Kuhn, 'China Goes One-on-One With the Net'; Drake, Kalathil and Boas, 'Dictatorships in the Digital Age'; K. Hartford, 'Cyberspace With Chinese Characteristics', *Current History*, September 2000, vol. 99, no. 638, pp. 255–62.
85 The kind of public discussion that takes place in electronic forums today over issues like homosexuality or AIDS would have been unthinkable some years ago. Although the number of taboos has decreased, however, subjects like the events surrounding the Tiananmen massacre of 3–4 June 1989 remain out of bounds.
86 Shapiro points out that when we consider the way a technology is used and the social environment in which it is deployed, taking into account the factors of

design, use, and environment, it becomes clear that the Internet might just as well suppress democracy as promote it ('The Internet', p. 15). However, while the sheer amount of information and multitude of opinions carried in cyberspace will not necessarily lead to more openness or tolerance of dissident views on its own, the multitude of Internet users with widely diverging interests and opinions may instead make it possible to create one's own 'global village', populated only by like-minded people and well secluded from the rest of the virtual world ('The Internet', p. 25).

4 Network convergence and bureaucratic turf wars

Junhua Zhang

According to the Tenth Five-Year Plan (2001–5),[1] the convergence of telecommunications, cable TV and computer networks is an important part of the drive to put the PRC at the forefront of the development of advanced information technology. Achieving this will be crucial for realising the hope of China's leaders that the telecommunications and ICT sectors can be used as engines for economic growth. As has been pointed out in previous chapters, tremendous progress has been claimed in these sectors and officials at the Ministry of Information Industry (MII) have reason to boast about their country's advanced technological position and the large scale of its broadband network, which acts as a stabilising influence on the economy.[2]

Such developments have been made possible over the past two decades by a sophisticated state-led development strategy that has enabled achievements like the completion of the nationwide fibre-optic backbone network two years ahead of schedule. Maintaining this rate of development, however, will be a strong test for the belief that ICTs can contribute to 'leapfrog' development.[3] This is because network convergence, the next generation of Internet development, will have two significant political impacts. First, the enhanced efficiency of services like telephony, TV and data transmission created by bringing networks together will deepen the threat to the CCP's monopoly on information. Second, the ideological tension that this generates is already becoming enmeshed in conflicts between powerful interest groups with stakes in the various digital networks concerned. This can clearly be seen in the turf war that is being fought between the MII and the State Administration of Radio, Film and Television (SARFT). Analysis of this conflict reveals how the prospect of network convergence raises a host of problems when China's existing political culture meets the technological opportunities and challenges generated by the latest stage of Internet development.

Prospects and challenges of convergence

China is geographically and institutionally a very large country that has a quite fragmented information infrastructure, each sector of which is usually dominated by one large player. This makes it especially important for the further development of the Internet that existing networks can be linked together in ways that allow for the more efficient use of resources, a reduction of costs for operators and end users, and an expansion of bandwidth. That the Chinese leadership recognizes the technical and economic importance of 'tri-net convergence' (*sanwang ronghe*) between telecommunications, cable TV and computer networks is evidenced by the promotion of this policy in the Tenth Five-Year Plan (2001–5), approved by the National People's Congress in March 2001. If this scheme is successful, then a rapid increase of Internet user numbers can be expected, with 10–15 per cent of the population brought online by 2005. Some observers in China are even more optimistic, suggesting that 60 per cent of households will be connected to the resulting broadband network, allowing around 200 million people to benefit from the modern Internet infrastructure.[4] If this target is met, then it will be no surprise if a new stage of social and economic development is indeed achieved and the Chinese language achieves a dominant status on the global Internet over the years to come.

Aside from the economic effect of 'tri-net convergence', however, far-reaching political consequences can also be expected from the connection of the cable TV network to the other two networks. In part, this is because the cable TV industry will have to operate more in line with the principles of the market economy, especially after accession to the WTO. This will shake the grip of the CCP on one of its last major ideological strongholds, leaving the 'special mission' of television considerably less tenable.[5] Such a development will further exacerbate the challenge to the ability of the authorities to disseminate propaganda and exert control over the supply of information that is already posed by the way in which the Internet requires Internet content providers (ICPs) and Internet service providers (ISPs) to play a large role in facilitating the assimilation of information and the provision of content to users.[6]

The challenge posed by network convergence to the ideological control of the CCP and to the plans of the government for a digital leapfrog is complicated still further by the way in which convergence requires new kinds of cooperation between bureaucratic organisations that have quite different aims. Cooperation will be essential, however, if the country's existing broadband network carriers are to decide on which common technology to adopt and how to go about acquiring and installing it.[7] Furthermore, 'convergence' means that there will have to be a very high degree of collaboration in the provision of technology, service, management and customer care between what have been three quite separate sectors until now. Yet rather than cooperation, the tendency so far has been for turf wars to break out. The most notorious of these to date has been between the MII and the SARFT. Unless such infighting between broadband network carriers can be resolved through a fundamental reorganisation of the existing communications industries and changes in the nature of Chinese business culture, the fibre-optic cables that have already been provided for many urban communities are likely to lie idle for some time.

To understand why a turf war between the MII and the SARFT has occurred, it is necessary to locate convergence within the broader context of the reform of China's telecommunications and IT sectors. This began with a first stage (1982–96) in which telecommunications companies were given more incentives to build and develop local infrastructures, and efforts were made to separate the functions of regulator and operator within what was then the Ministry of Posts and Telecommunications (MPT), which resulted in the rise of several operators by the early 1990s. Although most of these new operators worked on a basis of dubious legality, the government was willing to turn a blind eye because it welcomed the prospect of increased competition in the sector. This, however, turned out to be very limited due to the way in which the regulator gave preferential treatment to the original state monopoly.

The Asian financial crisis and fears of an economic slowdown stimulated a second stage of construction (1997–9), which saw the government launch an institutional and legal reform of the telecommunications and IT sectors involving a number of steps with far-reaching consequences. First, to meet the demand for technological convergence, the MII was created by merging the MPT (which oversaw network standards and access) with the Ministry of Electronic Industry (MEI) (which oversaw computers and software). The responsibility for postal administration and the telecommunications trunk line network was divested to other institutions. At the same time, the government introduced several measures to promote deregulation, including further breaking up the monopoly on telecommunications and granting newcomers preferential treatment with the aim of building up a more competitive industry that could offer a greater variety of services. Finally, a limited liberalisation of the market was introduced in tandem with the emergence of new competitors through an adjustment of service fees and charges. Such measures proved effective in insulating the development of the telecommunications and IT sectors against the impact of the Asian financial crisis.

The third stage of development began in 2000. In October of that year, with the worldwide Internet industry in a deep trough, the Fifth Plenum of the CCP Central Committee reaffirmed the belief in the possibility of achieving 'leapfrog' development through the promotion of ICTs by clearly emphasising the need for the further development of a modern information infrastructure, the enhancement of national broadband networks, and the convergence of telecom, video and data networks to improve the utilisation of the Internet.

One of the overall results of the programme of institutional reorganisation, market liberalisation and state directives was the emergence of no less than seven state-owned companies operating in the telecommunications and Internet industry, the technical specifications of which are listed in Table 4.1.[8] These figures show that, despite efforts by the government to create more competition, China Telecom has managed to remain the most powerful operator by maintaining the advantages that it inherited from the original monopoly that it held over telephone lines. Through its data communications

Table 4.1 Broadband specifications of seven licensed state-owned network carriers

Name of operator	Date of foundation	Broadband related infrastructure	International gateway Mbps	Broadband network related strategic outlook
China Telecommunications Group (China Telecom)	Restructured in January 2000. A further division will be completed by the end of 2002.	Eight vertical and eight horizontal fibre-optic backbone network with length of cable 200,000 km (July 2000). The nationwide digital microwave link totals 59,000 km. An advanced public data communications platform, part of it being ChinaNet, which has the largest coverage in China, as of 2001.	At the end of July 2000, the international outbound band of ChinaNet reached 711 MHz.	To complete a 15,000 km broadband network with 20 million users by 2005.
China United Telecommunications Corp. (China Unicom)	December 1993; restructured in 1999.	56,000 km of fibre-optic cable, reaching 250 cities as of 2001.	UNINET: 55 MHz (July 2000).	
China Netcom Corp. Ltd (China Netcom)	August 1999. SARFT is one of its major owners.	Joint pilot project: CNCNET – the nation's first backbone with 12,000 km fibre-optic cable and 40 Gbps overall bandwidth with 106 switching stations along a network linking 17 major cities (2001).	CNCNET: 377 MHz (July 2000).	To lay an additional 15,000 km of fibre-optic cable (2001).

Jitong Network	January 1994.	Owner of the Golden Bridge Project (CHINAGBN) – the national public economic information network.	CHINAGBN: 69 MHz (July 2000).	To construct the first national transmission backbone with the length of 4,700 km (2001).
China Mobile	December 1999.	CMNET (under construction), backbone linked with 28 provinces.	155 Mbps × 3.	
China Satellite Communications Group (ChinaSat)	1985; June 2000.	Unknown.	Unknown.	Unknown.
China Railway Telecom Corp. (China Railcom)	December 2000.	400,000 long-haul lines, trunk lines 120,000 km (fibre-optic cable 42,000 km), microwave range 4,590 km. The company opened its southwest loop line network on 21 November. The high-speed broadband network is 8,080 km long and of 80 Gb bandwidth (2001)		Planning to profit from concentrating on Internet and video transmission services in addition to traditional phone services.

Note
Military network excluded; China Telecom's ten northern provinces to be merged with Jitong Network and China Netcom in May 2002 as new China Netcom; China Telecom in south and west China to be restructured as new China Telecom.

arm, ChinaNet, it has now also become the leading Internet data centre, access provider and backbone proprietor in China. Survey data from the fourth quarter of 2000 from *iamasia* (Interactive Audience Measurement Asia) suggests that over 80 per cent of home Internet users access the Internet by using a China Telecom ISP. The company also provides more than 70 per cent of China's international Internet bandwidth.[9]

Although China Telecom faces more vigorous competition in the broadband access market, it is still the leading provider because most of its competitors are just beginning to launch these services. While leased lines are beginning to give way to more cost-effective alternatives, they remain the most common form of broadband connection, and China Telecom holds 90 per cent of the market. Building on its existing fixed-line network, China Telecom is moving to maintain its advantage by deploying integrated services digital network (ISDN) and asymmetric digital subscriber line (ADSL) services.

Regardless of the well-built national backbones that exist and the metropolitan networks that are controlled by a few alternative operators, achieving interconnectivity has been one of the weakest links in China's information infrastructure and a key reason why new entrants into the market have to struggle so much. In fact, connections between networks are so inadequate that, throughout 1999, traffic was often routed through a foreign country rather than directly between one domestic network and another. By early 2000 this situation had improved slightly, but even then only 10 Mbps of bandwidth connected Jitong to ChinaNet and 8 Mbps connected China's two academic networks, China Science and Technology Network (CSTNET) and China Education and Research Network (CERNET). This lack of interconnectivity was primarily the result of resistance from China Telecom, which sought to maintain its leadership by refusing or delaying interconnection to competing networks or charging exorbitant tariffs for those links once they had been opened. During 2000, removing these bottlenecks became a major priority for the MII. It proceeded by creating network access points (NAPs), first in Beijing and then in other cities, which would link all of China's network backbones together. In a compromise with China Telecom, which originally resisted participating in the NAPs at all, the former monopoly was given the task of managing these NAP facilities. The first of them, the China Internet Exchange (CNIX), opened in Beijing in March 2000.[10]

China Telecom continues to drag its heels in providing the free interconnection that the MII demands for CNIX as part of its general drive to improve interconnectivity, however. In Beijing, several rival operators have complained that China Telecom had only opened 10 Mbps of bandwidth to the MII-mandated NAP instead of the agreed-upon 155 Mbps, providing less than 10 per cent of what the MII wants. In Shanghai, China Telecom also refused to take part in a NAP organised by the local government. Moreover, representatives of rival backbones in Shanghai have expressed concern that China Telecom will open only limited bandwidth to provide access to the soon-to-be-opened NAP that is being sponsored there by the MII. As long as most of China's data communications traffic runs on China Telecom's network, the former monopolist's habit of limiting interconnectivity will hinder the ability of rival carriers to offer competitive services.[11]

China Telecom has also been able to maintain its monopolistic position because the MII has treated it as a favoured son until recently, for historical reasons.[12] Nevertheless, with the fostering of 'state-controlled' competition now an explicit goal of the MII, the anti-competitive practices that helped China Telecom to build its position in the marketplace are less and less tenable. A decision made in December 2001 by the MII to divide China Telecom into two parts again indicates that the ministry is determined to use its role as regulator to energise the market and prepare Chinese firms for the post WTO accession showdown. As competition has developed, however, the possibility for a turf war between the MII and the SARFT has arisen due to the prospect of convergence between the telecommunications network (under the administration of China Telecom and the MII) and the cable TV network over which the SARFT acts as both operator and regulator.

SARFT as a potential telecoms operator

The origins of the SARFT go back to 1954, when the Administration of Broadcasting Affairs (ABA) was created as an institution responsible for governance of the voice medium. After this was enlarged to cover multimedia in 1987, the organisation became the Ministry of Radio, Film and TV (MRFT). When a large-scale programme of administrative reform was carried out in 1998, the State Council at one point considered dissolving the MRFT by integrating it into different ministries. At the central level, however, this encountered severe resistance from the CCP Propaganda Department – sometimes called the 'invisible strong hand' behind the government. In the end, a compromise was reached by taking away the MRFT's ministerial status while retaining its regulatory control over broadcasting. Today's SARFT is the reincarnated version of the MRFT that emerged from this restructuring.

The reason why the SARFT has come into conflict with the MII lies largely in the way that both organisations were created during the 1998 restructuring of the bureaucracy. The MII, in acknowledgement of the growing inter-relationship between information industries and their potential power to drive economic growth, was made a super ministry with the authority to control all of China's information networks and play a direct role in creating relevant laws and regulations. This meant that as soon as the cable TV (CATV) network controlled by the SARFT began to be used for commercial purposes and was linked up with other telecommunications networks, the MII could claim the right to exert its regulatory functions over it. A turf war was thus inevitable, and made more intractable by the relationship of the SARFT with the CCP Propaganda Department, the closeness of which is revealed by the fact that the director of the former, Xu Guangchun, is at the same time deputy head of the latter.

Despite officially being under the direct supervision of the State Council, the SARFT is in fact deeply involved in overseeing the content of cable TV and other media on behalf of the CCP. It has for a long time played the role of both regulator and operator of the cable TV network. More specifically, it is responsible for:

- Approving the content of radio and TV programmes as well as films, overseeing film imports and stipulating the proportion of time to be allotted for foreign TV programmes, so that, in accordance with the requirements of the CCP's Propaganda Department, 'the Chinese people are not seeing programs that offend Chinese sensibilities or challenge the CCP's worldview'.
- Overseeing the operation of China Central Television (CCTV), the national TV network; approving the establishment of cable channels and the installation of head-ends in cable networks.
- Controlling access to satellite and cable networks as well as supervising their operation.

A special characteristic of the SARFT is the way in which it is structured into four layers of central, provincial, district and county management. This has been particularly important for its relationship with the bottom-up emergence of the cable TV network since the late 1980s, which has been constructed and regulated mainly by local governments, with many provinces establishing their own networks. The popularity of these networks has been boosted by the arrival of new satellite broadcasting and foreign cable services, such as ESPN, and the large profits that have been earned by suppliers of hybrid fibre-coaxial cable and multipoint microwave distribution systems for the building of provincial and municipal systems. The resulting boom in the sale of cable TVs has made the SARFT a great potential broadband operator that others can only envy. It controls a network of over 2.4 million kilometres with 0.3 million kilometres of fibre-optic cable, and has at its disposal 14,000 kilometres of national trunk fibre connecting 22 provincial HFC networks. Some 90 per cent of the existing network consists of one-way cables, with 70 per cent of cabled households using 550 Mhz–750 Mhz HFC technology.[13] With such advantages, once the SARFT has upgraded to a high-speed two-way infrastructure it will be well positioned to offer high-speed Internet access, something that has proved to be a profitable new revenue stream for cable operators and that has huge potential in the Chinese context, as shown by the international comparison in Table 4.2.

While the figure for telephone subscribers has also dramatically increased in recent years, with 144.4 million fixed-line and 70.9 million mobile subscribers at the end of 2000,[14] the development of CATV draws more attention from players interested in the deployment of broadband. By the end of July 2001 TV penetration had reached 93.5 per cent of households and CATV penetration about 30 per cent. China's major cities have nearly 100

Table 4.2 International comparison of household penetration of networks (1998)

	China (%)	USA (%)	India (%)
TV penetration	84	220	24
Telephone penetration	11	93	6
Computer penetration	1.20	35	0.80
Cable TV penetration	23	68	9

Source: SARFT, 2001.

per cent cable penetration. The total length of CATV network amounts to 2.6 million kilometres.[15] Table 4.3 indicates the rapid rate of increase of CATV subscribers in the 1990s.

Realising the commercial significance of convergence, the central administration of the SARFT has vowed to establish a modern broadband network in China in ten years by modifying the current TV-oriented network into a two-way broadband net. The first stage of construction is planned to end in 2005, completing the tough job of bringing distributed networks together while simultaneously expanding the current 90 million subscribers to a figure of 150 million. The total number of subscribers is expected to reach 200 million during a second stage of development scheduled to finish in 2010, with hybrid fibre-coaxial cable being the main access technology deployed on the network.

Such initiatives can also be seen as part of an effort by the central management of the SARFT to regain control over its own organisation following the weakening of its power to act as an operator that took place in the 1990s, with the aim of clawing back the lion's share of the revenues brought in by its local bureaus. In view of the strong trend towards regionalism, however, leaders of the SARFT have legions of obstacles to overcome if they are to gain control over the 4,000 or so cable networks that have been developed from the bottom up and around 1,300 cable TV stations. More and more local branches are trying to steer away from central supervision, unwilling to give up what they have invested in and built up, knowing that they will not be adequately rewarded if a fusion does occur. In fact, the existing structure of the networks and their stockholders has become so complex that a smooth realisation of mergers will be difficult even if all sides are willing to cooperate and even though the SARFT has secured up to RMB 40 billion in bank loans with which to buy up cable TV stations. Obviously the SARFT will have

Table 4.3 Increase of cable TV subscriber numbers (million households)

1990	1997	1999	July 2001
13	56.5	80	84.76

Source: Screen Digest 1998 www.screendigest.com; SARFT, 2001.

to employ 'administrative methods' to enforce the local 'warlords' to accept its offers.

In terms of its institutional structure, the SARFT has also been trying over several years to introduce the policy of separating its functions as regulator and operator. This, however, is not so much to meet the requirement of the State Council and the MII, as it is to prepare for large-scale participation in the competition for the broadband network and to make itself fit to run a digital TV business effectively. An early step in these efforts was the creation of the China Data Broadcasting Centre in 1994. This is expected to become the scaffolding of the China Data Broadcasting Group, the purpose of which is to engage in the business of data broadcasting in the PRC. In 1999 the SARFT took another step forward by establishing the China Broadcasting Information Network Centre to take on the specific task of handling cable-network expansion. In December 2001 the China Radio, Film and Television Group (CRFTG) was formed as a new media holding company to gather SARFT's assets under one roof – including China Central Television (CCTV), China National Radio, China Radio International, China Film Group, China Radio and Television Transmission Network, and China Radio and Television Website. This company is to be run on a *de facto* semi-commercial basis with a sweeping range of activities ranging from programme production to publications and the export of CCTV programmes as well as arts performances, advertising and real estate. The group will also be responsible for reforming the national radio and television networks by setting up a nationwide transmission system that will fully utilise high technology, cultivate and explore new areas for economic growth, and expand into the international market to bring the voice of China to the world.

The special status of the SARFT, however, means that it is doubtful whether the organisation can really be successful in separating its business operations from its administrative functions. The inseparable link between politics and business when it comes to media control is underlined by the fact that the CRFTG is chaired by Xu Guangchun and the SARFT holds at least 51 per cent of the group's shares. The SARFT thus occupies an unassailable position so long as the CCP is not ready to give up its monopoly on power. Since TV has become the most popular medium in China, the very existence of the CCP depends very much on how it is able to manipulate this information resource and its related infrastructure, regardless of the fact that such manipulation has become quite different from what it was twenty years ago. The SARFT thus has an extremely powerful protectorate to appeal to whenever it runs into business risks.

Two battles between the MII and the SARFT

Conflict between the MII and the SARFT, then, is made more likely by the fact that these two organisations are pursuing quite different goals. For the MII it is vital to maintain the rapid growth of the telecommunications and

IT sectors in order to warrant a high level of GDP, while it also benefits a great deal from businesses run by its former 'sons' like China Telecom or even far-related 'family members' like Unicom or other IPSs. However, since it is charged by the State Council with regulating the telecommunications market, it has to play a neutral role when it deals with competing operators. The SARFT, by contrast, always has to consider the political consequences of any kind of practice that might benefit or harm the Party, or at least appear to be doing so if it is to stay on-side with the CCP's Propaganda Department and maintain its special status. In reality, however, the way in which the two organisations work shows that both – and especially the SARFT – endeavour to gain more terrain for their own business. In this respect, the SARFT is not satisfied to remain solely a propaganda-oriented institution and turns out to be just the opposite in many cases.

To understand the significance of these developments the positions of both organisations can be illustrated by looking at the two rounds of conflict that have taken place between them since they emerged from the ministerial reforms of 1998 and decided to seize a slice of the telecommunications market. The MII set about this mainly through working with China Telecom, which had become active in concluding contracts with different operators who were entitled to use the cable TV network. The SARFT, however, could claim to have overall jurisdiction over the country's radio and TV industries based on the 'Measures for the Administration of Radio and Television' that had been issued in 1997. It thus extended its authority over the construction and operation of radio and TV transmission networks, as well as over television programming. The overall restructuring that took place in 1998, however, meant that the administrative jurisdiction of the SARFT was in fact left somewhat hazy. Up until 2000, under the leadership of Tian Congming, this opened up possibilities for taking advantage of legislative grey areas in order to participate in the telecommunications business and start providing Internet access and IP-backed telephony. A significant concrete step was taken to further Tian's ambitions to gain a foothold in the potentially lucrative Internet sector by exploiting legislative ambiguity when the telecommunications company Netcom was established, in which the SARFT is a shareholder. With the approval of the State Council in 1999, Netcom was allowed to use the SARFT's local networks and to further develop a national fibre-optic backbone.

Both the MII and the SARFT have developed theoretical arguments to try to legitimate their endeavours. On one side, articles such as Wang Xiaoquan's 'Development Strategy of China's Telecommunications Industry', published in the March 1998 issue of *Industry Forum* (*Chanye luntan*),[16] argues that China can only compete against the world's industrial giants if the government maintains control and actively uses its control to merge the three kinds of information networks (*sanwang heyi*) to create an integrated mega-network.[17] Yet, this mega-network should ensure the dominant status of China Telecom because of its existing nationwide infrastructure of eight horizontal and eight vertical fibre optic backbone network, which makes constructing a new network by the SARFT somewhat unnecessary.[18] Acting on behalf of the SARFT, Fang Yihong published an essay three months later in order to respond to Wang's challenge. While he agreed with Wang's view on the necessity of opening up the telecommunications market, Fang argued that this would be best achieved by allowing the telecommunications network and cable TV network to compete against each other. Only through such sound competition would it be possible to build a mega-network based on the new stage of cable TV broadband technology, he proposed.[19]

While the turf war was being waged verbally at the highest levels, local companies had more aggressive ways of expanding their market inch by inch. From 1994 to 1999 in the province of Hunan, China Telecom and the SARFT were literally in a state of war, with dozens of people from both sides shot dead and about 400 suffering injuries during tussles.[20] In order to end this battle, the State Council issued a directive banning the municipal and provincial organisations of China Telecom and the SARFT from taking actions against each other.[21] In other words, both sides were ordered by the central government to restrict their business to their own terrain, with telecommunications entities told not to engage in radio and television transmission, and radio and television entities not allowed to step into the telecommunications business.

The second round of struggle began after Xu Guangchun was nominated as director of the SARFT. The specific cause of tension was a new round of regulations issued by the MII, according to which it claimed jurisdiction and control over transmission networks once they began to be exploited for commercial purposes. The MII's power in this respect has since been codified in the Telecommunications Regulations that were issued in 2000,[22] according to which the ministry has authority over all 'telecommunications activities'. Such activities are very broadly defined to include any activity involving the use of a cable or wireless electromagnetic system or electro-optical system to transmit, send or receive voice, text, data, graphics or any other form of information. This clearly covers what has traditionally been considered television and radio transmission, always the province of the SARFT. However, the SARFT has remained unwilling to give up its control over the TV and radio transmission networks, which represent valuable infrastructure not only for the provision of telecommunications services but for broadcasting as well.[23]

There is an obvious inconsistency, then, between the Telecommunications Regulations of 2000 and the State Council's directive of 1999. The SARFT interpreted the directive liberally and made use of it to resist the MII's jurisdiction over cable networks, as well as to preclude telecoms service providers from transmitting programmes over telecommunications networks. At the same time, though, it continues to promote its own broadband business in hidden ways. Yet, fearing political pressure from the CCP, the State Council has not yet attempted to resolve this inconsistency between the two edicts. The law-makers in China are quite aware that the ideological basis for SARFT's inflexible position is grounded in the belief that the proper function for broadcasting is to disseminate CCP and government information. Moreover,

China has not yet committed itself to opening radio and cable television to foreign participation as part of its accession to the WTO, although it has made such a promise in respect of telecommunications services.

Far from satisfied with such a state of affairs, in July 2001 the MII attempted to exert its jurisdictional authority by issuing a regulation concerning the conditions for investment in the telecommunications business. According to this, if a licensed telecommunications carrier has fewer than 51 per cent of the shares in any joint venture set up by basic telecommunications operators and other domestic state-owned firms, then the joint venture should apply to the MII for a licence before it engages in any telecommunications business. Although the MII claims that this rule does not target any individual carrier, it is widely believed that China Netcom will be the first target. Indeed, it may be that the SARFT, as one of China Netcom's main owners, is the MII's real target. This is because the SARFT, the entertainment programme regulator, has provided investment for China Netcom to form a number of joint ventures with local broadcasting companies in various provinces and cities, with less than 51 per cent equity. These joint ventures, in cities like Hangzhou, Qingdao and Chongqing, operate basic telecommunications businesses such as data transmission, network bandwidth release and Internet access. Under the new regulations, therefore, most of them should apply to the MII for licences.

At the same time, being eager to see the realisation of 'tri-net convergence,' the MII held out an olive branch to the SARFT in October 2001, saying that it encouraged the equal entry of telecommunications and broadcasting companies into the market. In response to the MII's call for a reciprocal opening of the telecommunications market and the radio and TV transmission market, top officials at the SARFT suggested a unilateral opening for the radio and TV network operators to enter into the telecommunications sector. Chen Xiaoning, director of the SARFT Information Network Centre, made a very interesting point when he said that the telecommunications networks are like public parks that should be open to all sightseers, while the SARFT's are like military camps that might only be open for visits from time to time. Doubtless, officials at the MII do not agree with Chen's view and it remains to be seen which organisation's view will prevail concerning who should regulate China's television and radio transmission network.[24]

Up to the present, then, the question of how 'tri-net convergence' will be accomplished still remains open. But since the central government is keen to promote a national broadband network, a solution to the conflict between the MII and the SARFT will have to be found. It seems, however, that the existing super-governmental agency known as the 'National Informatization Leading Group', headed by Zeng Peiyan, minister of the State Development Planning Commission (SDPC), is not capable of fulfilling this task. It is for this reason that Premier Zhu Rongji set up an 'Office for Information Industry' at the beginning of 2001, with himself as chairman. If the aim of this new organ is to control turf wars, though, it will still take some time before the golden mean between these battling interest groups can be found.

Three major contradictions

China is going through a period of radical transformation during which social change is determined by innumerable contradictions. The case of the turf war between the MII and the SARFT illustrates at least three of the most basic of these contradictions, namely political versus economic imperatives, the rule of law versus cynicism towards any kind of regulatory framework, and macroeconomic planning versus the activities and management of interest groups. Concerning the first of these, it should be appreciated that not all of the SARFT's arguments for protection are baseless, bearing in mind that even the Western liberal-democracies have introduced regulations to protect their own TV networks. The dilemma faced by the Chinese leadership is particularly acute, though, because it wants to enhance economic performance by removing obstacles that are in fact the very means upon which it bases its rule. This problem is particularly apparent with reference to the development of a converged broadband network, which may cause the national economy to forge ahead in a major way but will also bring about a certain shrinkage of control over the ideological terrain in the process, if sufficient content to satisfy economic needs and the desires of end users is to be provided.

The present credo of the Chinese leadership, however, is that China should have both economic progress and political control through propaganda. As Jiang Zemin himself has said, 'under the conditions of a market economy our news agencies should run propaganda as well as business'.[25] In reality, though, many top officials are quite aware that 'tri-net convergence' can be achieved only if the SARFT is ready to give up some of its influence. The prevailing tendency will certainly be to move in this direction so long as the national leadership is determined to realise the ambitions it has codified in the Tenth Five-Year Plan. The question remains, though, as to what formula can be found to justify the shrinkage of the SARFT's ideological power while the CCP is also able to maintain its assertion that it has a monopoly on political life. In the long run, therefore, the progressive developments presently underway in China's ICT sectors can be regarded as a journey into an uncertain future for the CCP.

The second contradiction concerns the sincerity of big players when it comes to observing regulations. Constructing China's national economy needs a transparent legal framework. With specific reference to network convergence, it has been pointed out that 'Without a transparent regulatory agency it is extremely difficult to negotiate and enforce terms of interconnection between carriers'.[26] While great efforts to improve transparency have been made over recent decades, and accession to the WTO will accelerate the pace at which the rule of law is strengthened, the dispute between the MII and the SARFT illustrates all too well the ways in which the political culture and system in China can produce a certain degree of cynicism towards laws and regulations, a general phenomenon that has long been noted by Western social scientists.[27] This is partly due to the way in which laws and regulations in

China tend to have a very short-term character and often lag behind reality, something especially true of so-called 'red-head' directives that claim to function with binding legal force. It is also due to the way in which the one-party political system itself mitigates against the effective implementation of regulations, as shown by the ability of the SARFT to rely on its political background to circumvent the Telecommunications Regulations. The lack of a clear separation between regulators and operators, although the central government has strived to achieve such a separation for more than a decade, also does little to alleviate cynicism. In this respect, the movement from the planned economy to the market system is obviously a transition that is still in process.

The third contradiction lies in the inappropriate relationship between central government policy-makers and the major actors supposed to implement the policies they make. One result of the devolution of power to ministries and local authorities that has taken place over the past two decades has been the formation of numerous interest groups, with organisations like the MII and the SARFT being particularly difficult to tame. While the phenomenon of lobbies is commonplace in many other countries, however, a real challenge for the Chinese leadership can be created when pressures are exerted by ministerial and local sources. This will be even more the case for the fourth generation of leaders that will emerge after the Sixteenth CCP Congress in 2002. The lack of charisma of the new power holders means that the central government will have to create a better legal and institutional framework if it is to maintain the authority it needs to meet the targets set for the national economy. Because regulatory structures can be opposed by big players with strong vested interests, however, bargains between the central government and the ministries will have to be struck. However, there is little point in striking such bargains if the agreements reached are not maintained by all the parties concerned. Moreover, being able to forecast developments and come up with a suitable legal framework that can break up monopolies and create a level playing field is something that requires a considerable degree of sophistication on the part of policy-makers and political leaders.

In spite of the mass of problems that the central government faces concerning 'tri-net convergence', however, the ability of Chinese policy-makers to learn through a process of trial and error should not be underestimated. In fact, throughout the period of 'reform and opening' initiated by Deng Xiaoping, new ideas have often been introduced from the outside world through ingenious devices such as the 'special economic zones'. The same kind of method can be seen at work with reference to ICTs in a document issued by the MII in July 2001, according to which thirteen cities are permitted to operate broadband businesses, with the SARFT involved as something of a hidden player. Shanghai, for example, is allowed to provide citizens in several of its districts with a qualified broadband service, in which the SARFT has been given official permission to cooperate.[28] Even some stations belonging to local cable TV networks have already begun to experiment with their own 'tri-net convergence' in cities and regions such as Chongqing, Qingdao, Jinan, Nankai and Foshan.[29]

At the national level, in 2001 the Chinese Academy of Sciences (CAS) initiated a pilot project named China Advanced Info-Optical Network (CAINET), under the auspices of Premier Zhu Rongji. This will have far-reaching consequences for the Internet industry, in which the SARFT, the Ministry of Railways and the Shanghai Municipal Government are already taking part. Such schemes indicate that there are no *de facto* obstacles stopping the SARFT from engaging in the Internet business. Moreover, although CAINET has only been partly realised so far, its implementation has already narrowed the lead of the developed countries to a gap of two years in this field, making China one of the world's most advanced nations in network science and technology.[30] The experiences of such pioneering projects will provide the central government with much valuable knowledge for the further implementation of its broadband plans. Throughout the period of 'reform and opening' action has often preceded the introduction of new regulations and has frequently been used for the benefit of good decision-making. However, it should be pointed out that when trials are being conducted, it is difficult for non-favourite business players, including foreign companies, to operate in areas that lie outside the limits of the designated 'special zones'.

Concluding remarks

As long as the CCP Propaganda Department and the SARFT, its media extension, exist, the single-party system in the PRC will remain a hindrance to plans to make ICTs an engine for economic growth. This is because the CCP views control over the media as vital for its survival at a time when economic liberalisation is bringing a host of new challenges, including the threat of civil unrest as foreign competition throws millions out of work. However, the national leadership is also reconciled to the fact that its ideological strongholds will have to make concessions if the kind of development in the telecommunications and IT sectors that can promote economic growth is to be feasible. Moreover, as a December 2001 Propaganda Department document regarding permission for foreign investment in the media sectors shows, there are strong incentives even among organisations like the SARFT to begin to enrich the media market.[31] To be sure, the introduction of foreign capital does not in itself mean that the CCP is deliberately abandoning political control. Space will be created, however, in which the market and market mechanisms can grow as interest groups like the SARFT strive to achieve greater profits. As a result, the influence of the Propaganda Department will tend to fade.

Within the global context, one could argue that the problems of 'tri-net convergence' faced by China's central government are common to some degree to all countries that are struggling to define the content of a broadband network and which authorities should be responsible for it as the Internet heads into a new stage of technological development. Like other countries, China still needs time to create an appropriate legal framework to deal with these issues. Moreover, the strong belief amongst the leadership in the possibility of achieving 'leapfrog development' via telecommunications and the Internet industry will force them to find a solution to these problems in order to break out of what could become one of the bottlenecks of further development in the coming years.[32] If such problems are solved, then a very different landscape for the Internet in China and a new constellation of competition among operators and providers will emerge. It is impossible to predict whether such a process will produce negative results, but positive effects such as a reduction of the digital divide and a general promotion of information technology can be expected.

Last but not least, the conflict between the MII and the SARFT has been highlighted here as a typical example of a specific kind of problem that afflicts China's development, namely the tension that builds up between big economic players with a stake in the political status quo. Other actors in China's ICT sector, such as China Railcom and China Mobile, are also engaged in conflict with the aforementioned competitors. The difference between such battles and the tensions between authorities and operators in consolidated liberal-democracies is that the contestants in the latter do not need to justify their actions by appealing to ideological rhetoric because there exists a relatively clear demarcation between what is permitted and what is not. In China, however, it will take several years at least, and probably decades, before telecommunications carriers themselves (including cable TV) can lease the network from each other on the basis of clearly defined legal ownership, rather than looking to the state, the ministry or the Propaganda Department. Of course, when that time comes, it will finally be possible to say goodbye to the one-party state.

Notes

1 State Council, *Shiwu jihua gangyao quanwen* (*Complete Text of the Outline of the Tenth Five Year Plan*). Online. Available HTTP: <http://www.chinaemb.or.kr/chn/9272.html> (accessed 10 April 2002). See also '*Zhongguo xinxi chanye shiwu fazhan guihua*' ('Development Programme of Chinese Information Industry in the Tenth Five Plan'), Beijing: *Renmin chubanshe*, 2002.
2 Han Xia, '*Zhashi tuijin woguo de kuandai jianshe he yingyong*' ('Forging the Construction and Application of China's Broadband'), 23 August 2001. Online. Available HTTP: <http://www.mii.gov.cn> (accessed 20 December 2001).
3 Yang Haifeng, '*Zhuanjia lun xiayidai wangluo*' ('Experts on the Next Generation of the Internet'), *Hulianwang Shijie*, 2002, no. 1, p. 62.
4 Zhen Yong, '*Sanwang ronghe caidian deshi, kuandai qudong dianzi shangwu*' ('Colour-TV Will Benefit from "Tri-Network Convergence", Broadband Will Forge e-commerce via TV'), *Beijing wanbao*, 6 April 2001, p. 5.
5 Chen Xiaoning, '*Lun youxian yu xinmeiti de guanxi*' ('On the Relationship Between Cable TV and New Media'), part 5, 2 December 2000. Online. Available HTTP: <http://www.sarft.com> (accessed 2 February 2001).

6 Liang Ping, *'Tuijin woguo sanwang ronghe jincheng de sikao'* ('Considerations about How to Accelerate the Process of China's Tri-Net-Convergence'), *Hulianwang shijie*, 2002, no. 4, p. 30.
7 Liang Ping, *'Tuijin woguo sanwang'*, p. 29.
8 On 17 May 2002 the number of operators will be reduced from seven to six when the part of China Telecom that covers ten of the northern provinces will be merged with Jitong Network and part of China Netcom to form a new China Netcom. The former China Telecom in south and west China will be reorganised into a new China Telecom.
9 Xiao Hongzhi, *'Dianxin fuwu qushi tansuo – dalu dianxun shichang xiankuan pingxi'* ('Probing the Trend of the Telecommunications Sector – A Critical Assessment of the Telecommunications Market in Mainland China'), 8 May 2001. Online. Available HTTP: <http://www.find.org.tw/0105/trend/0105_trend_disp.asp?trend_id=1158> (accessed 3 December 2001).
10 T. Dean, 'Telecommunications: The Data Communications Market Opens Up'. Online. Available HTTP: <http://www.chinaonline.com> (accessed 28 December 2001).
11 T. Dean, 'Telecommunications: The Data Communications Market Opens Up'.
12 M. Mueller and Zixiang Tan, *China in the Information Age: Telecommunications and the Dilemmas of Reform*, Westport, Connecticut and London: Praeger Publishers, 1997.
13 David Tao, 'Asia Cable TV – A Broadband Service and Content Provider in PRC'. Online. Available HTTP: <http://www.asiacabletv.com> (accessed 2 January 2001). For an explanation of HFC technology please see: http://www.c114.net/technic/technicread.asp?articleid=3812&boardcode=jl_catv (accessed 3 May 2002).
14 Sunray Liu 'China Stands at Broadband's Gate', *EE Times*, 23 April 2001. Online. Available HTTP: <http://www.eetimes.com/story/OEG20010423S0122> (accessed 29 December 2001).
15 Details can be found on the SARFT Website at <http://www.sarft.com>.
16 Wang Xiaoquan, *'Zhongguo dianxun chanye de fazhan zhanlüe'* ('Development Strategy of China's Telecommunications Industry'), *Chanye luntan*, March 1998, cf. Zhou Qiren, *Shuwang jingzheng (Competition Among Telecommunication Networks)*, Beijing: Sanlian shudian, 2001, p. 3.
17 There is a difference between *'Sanwang heyi'* (making three networks into one) and *'Sanwang ronghe'* (tri-net convergence). While the former phrase implies the dominant role of either China Telecom or the cable television network, the latter just indicates the cooperative utilisation of various networks according to market demand.
18 Zhou Qiren, *Shuwang jingzheng*, p.3.
19 Zhou Qiren, *Shuwang*, p. 3.
20 *'Hunan youxian wangluo dazhan shangwang bairen'* ('More than 100 Injured or Dead in Hunan Cable Network Turf War'). Online. Available HTTP: <http://www.cctv.com/specials/dianxin/main/dianxin63.html> (accessed 21 December 2001).
21 Ministry of Information Industry and SARFT, *'Guanyu jiaqiang guangbo dianshi youxian wangluo jianshe guanli de yijian'* ('Opinions on Strengthening the Management of the Construction of the Broadcasting and TV Cable Network'), 13 September 1999, *Zhonghua renmin gongheguo guowuyuan gongbao*, 1999, no. 35, pp. 1573–5, document 82 approved by the State Council Office, 1999.
22 Ministry of Information Industry, *'Zhonghua renmin gongheguo dianxin tiaoli'* ('Telecommunications Regulations of the People's Republic of China'), 25

September 2000. Online. Available HTTP: <http://www.mii.gov.cn/news2000/1013_1.htm> (accessed 2 December 2000).
23 '*Guandian: guandianwang he dianxinwang shi "tongzhi yigou"*' ('Viewpoint: Cable Network and Telecoms Network are of the Same Quality but not of the Same Structure'). Online. Available HTTP: <http://tech.sina.com.cn/it/t/63482.shmtl> (accessed 27 August 2001).
24 Zhang Dongcao, '*Dianxinye mianlin bianju: youxian dianshi jiang dapo dianxin longduan*' ('Telecommunications Sector is Facing Change: Cable Network will Break Telecommunications Monopoly'), 12 July 2001. Online. Available HTTP: <http://tech.sina.com.cn/it/t/75514.shtml> (accessed 20 July 2001); see also Chen Xiaoning, '*Lun youxian yu xinmeiti de guanxi*'.
25 Chen Yichong, '*Zhongguo chuanmei gu huo zai ziben yu zhengce jiafen zhong*' ('The Stock of China's Media is Struggling Between Capital and Policies'). Online. Available HTTP: <http://tech.sina.com.cn/it/75784.shtml> (accessed 25 July 2001).
26 K. Zita, 'LMDS and Broadband Local Networks for Asia', 9 July 1999. Online. Available HTTP: <http://www.vii.org/papers/ptc97.htm> (accessed 30 December 2001).
27 L.W. Pye, *The Mandarin and the Cadre: China's Political Cultures*, Ann Arbor, Michigan: Center for Chinese Studies, University of Michigan, 1988, p. 44.
28 In MII and SARFT, '*Guanyu jiaqiang guangbo dianshi youxian wangluo jianshe guanli de yijian*'. Shanghai was entitled to use all available networks to develop broadband service. See Liang Ping, '*Tuijin woguo sanwang ronghe jincheng de sikao*', p. 30.
29 '"*Chuanmei diyi gu*" *de fan nao*' ('Plague of "First Stock of Media"'), 10 July 2001. Online. Available HTTP: <http://tech.sina.com.cn/it/t/72173.shtml> (accessed 11 July 2001); Zhao Jin and Xu Hongzhou, *Xinxi Zhongguo* (*Information China*), Beijing: Jingji kexue chubanshe, 2001, pp. 37–50.
30 Zhou Qiren, *Shuwang jingzheng*, p. 106.
31 CCP Propaganda Department, Document No. 17, 2001.
32 See Zhang Junhua, 'Will the Government "Serve the People"? The Development of Chinese E-government', *New Media and Society*, vol. 4, no. 2, June 2002, pp. 163–84.

5 (Re-)Imagining 'Greater China'
Silicon Valley and the strategy of siliconization

Ngai-Ling Sum

The chapters in this book leave little doubt that China has been caught up in a global 'Internet rush' that is primarily concerned with the practical application and exploitation of ICTs. Yet the emergence of an information age that includes a knowledge-based economy and a learning society is also prompting the widespread reorganization of economic, political, and social relations not only within national societies but across borders too. A central movement that is promoting this tendency is what might be called the global 'Silicon Valley Wave'. This is a trend that tries to harness the benefits of broader changes in the relations of production by emulating the original Silicon Valley in California. It can be seen throughout the world in projects such as Singapore's 'Intelligent Island', Malaysia's 'Multimedia Corridor', Scotland's 'Silicon Glen' and Israel's 'Silicon Wadi'. Within the PRC, it can be seen for example in the concentration of high-tech research and development at Beijing's Zhongguancun science and technology park.

This chapter will focus on the ways in which the imagination of a Silicon Valley model is impacting on the formation of the regional economy of 'Greater China', an identity that began to emerge after the opening up of the PRC in 1978. This encouraged Hong Kong's manufacturers and traders to start to play a role in consolidating a new cross-border division of labour by investing in southern China, soon to be followed by their counterparts from Taiwan.[1] Such developments led policy-makers in the region, such as Singapore's Goh Chok Tong, to see the emergence of 'Greater China' as a paradigmatic case of cross-border economic cooperation. Terms such as 'natural economic territories', 'growth triangles' and 'cross-border regions'[2] capture this kind of cross-border phenomenon in regional studies, geography, and international studies, as well as in policy-making circles. Central to such concepts is the proposal that changes brought about by globalization lead state, quasi-state and non-state actors at the local, national, trans-local, regional, and global levels to become involved in the development of new governance regimes that are not solely tied to nation-states. The distinctive feature of such regimes is the way in which economic activity is mediated in and through network-type modes of coordination, rather than through either more traditional arms-length market forces, or top-down planning and direction by the state.[3]

As for the original Silicon Valley, this has been described as a 'high-tech district/cluster' in California with a dense concentration of Internet-related and other computer-based activities.[4] It is a habitat that includes people, firms, institutions and their networks and modes of interaction.[5] Such networks have been built by venture capitalists, specialized professional firms, universities, and research centres. There is a pool of high-tech workers who generally embrace an entrepreneurial ethic and are willing to engage in a high level of collaboration. Since 1980, some 10,000 firms have started up in the region, many of which have become large listed companies on the New York Stock Exchange (NYSE) and the Nasdaq, such as Intel, Cisco Systems and Hewlett-Packard.

When we look closely at the way in which the imagination of the Silicon Valley model has interacted with the formation of 'Greater China', it becomes clear that it has been used by private and public actors try to try to reorganize social relations through a variety of tactics that constitute a strategy of 'siliconization'. Central to this is the construction of multiple 'Silicon Valleys' as new objects of economic, political, and cultural governance. Once imagined, such hi-tech clusters can then be used to bring about basic changes in practices and social relations in order to consolidate a new regional growth dynamic. It will be argued below, however, that even if the consolidation of such a new cross-border mode of growth and governance is successful, then it will still reflect the significant geo-economic and geo-political conflicts and struggles that characterize any particular region.

Siliconization and the post-crisis remaking of 'Greater China'

Public and private actors throughout the world have come to admire Silicon Valley as a model for development and a possible basis for their own future economic success.[6] Since the Asian financial crisis began in 1997, such a vision has become particularly attractive in Hong Kong and Shenzhen. It has also taken on an additional significance in Taiwan, although it had been advocated there for some years already. Powerful forces in the region can thus be seen to be actively reorienting and re-imagining economic formations through a strategy of 'siliconization', which involves the following key elements:

- The privileging of 'information and communication technology' (ICT) and the 'Silicon Valley' discourse that is promoted by diverse private and public actors.
- The construction of 'Silicon Valley' as a new form of knowledge and new object(ive) of 'future growth' that involves new global–regional–national–local linkages in the 'information age'.
- The assemblage of techno-economic practices by diverse actors in the hope of stabilizing an emerging growth regime based on the 'new economy'.

- The mediation and consolidation of this mode of governance through both cooperation and competition as well as through the contestation of identities and practices.

This strategy of siliconization in the post-crisis period can be discerned in official/public discourses throughout 'Greater China'. In Mainland China, the close linkage of the discourse of Silicon Valley with 'high technology' and the 'Asian crisis' can be seen in remarks made by PRC Premier Zhu Rongji during a visit to the southern city of Shenzhen in October 1999 to attend the first China High-Tech Fair, when he pledged that, 'We must learn the lesson of the Asian financial crisis. Only when high technology is developed can we avoid the phenomenon of a "bubble economy"'.[7]

Zhu's optimistic linkage of 'high technology' with 'bubble avoidance' resonates with the sentiments of President Jiang Zemin's policy of 'reinvigorating the nation through technology' (such as information technology, space science and new energy) which goes as far back as the 1988 Torch Plan, promoted by the State Department of Science and Technology at that time. At the local government level, Guangdong Province sought to implement this same plan by constructing its own Silicon Valley in the Pearl River Delta. This project involved a total investment of USD 11 bn and concentrated on research and development, training and education, trade and industry and hi-tech exhibitions. The location of the first state-level hi-tech fair in Shenzhen, and Zhu Rongji's participation, symbolized the status of the city as a hi-tech zone. This was reinforced when Shenzhen mayor Li Zibin visited the Guangdong Technological Innovation Symposium in March 2000 and announced that his city would focus on the 'high and new technology sector, especially information products based on the Internet and a digital Valley'. At the Information Technology Working Conference in Shenzhen, Li again specifically stressed his ambition for the information industry to form a solid foundation for Shenzhen to become 'China's Silicon Valley', with an IT output value expected to reach USD 16.9 bn by the end of the year 2000.[8]

The 'hi-tech' Silicon Valley model can also be found in post-1997 Hong Kong. In March 1998, Hong Kong Chief Executive Tung Chee-Hwa set up the Commission on Innovation and Technology to examine the possibility of turning Hong Kong into a centre of innovation and technology; its first report provided the background for him to link 'hi-tech' with the 'Asian crisis'. In his first policy speech, delivered in October 1998, Tung thus remarked:

> To help our economy respond to the changes I have described [Asian crisis], our strategy will be to focus on increasing the diversity of the economy by creating conditions for growth in sectors with a high value-added element, in particular in those industries which place importance on high technology and multi-media applications.[9]

Hong Kong Financial Secretary, Donald Tsang, translated this push for high technology into more concrete projects when he delivered his budget speech the following year. According to Tsang:

> There is no question that, for Hong Kong to meet the challenges of the 21st Century, it must adapt to the new forces of the Information Age. Technological advances such as digitalisation and broadband networks are introducing new ways of doing business, transforming traditional markets and altering existing competitive advantages.
>
> To respond to these mega trends ... the Government proposes to develop a 'Cyberport' in Hong Kong. The Cyberport will provide the essential infrastructure for the formation of a strategic cluster of information services companies. These companies would specialize in the development of services and multi-media content to support businesses and industries.[10]

These 'high-technology Silicon Valley' discourses in Mainland China and Hong Kong resonate with similar visions that have emerged in Taiwan since the 1980s. It was then that the 'Silicon Valley' idea motivated the Hsinchu science-based industrial park and government plans to attract Silicon Valley 'returnees' through tax breaks and other economic incentives and through organizations such as Monte Jade.[11] Almost half the companies in the Hsinchu park in 1997 were started by United States-educated engineers. They concentrated on OEM (original equipment manufacturing) and networks of producers manufactured almost 40 per cent of notebook PCs sold in the world along with two-thirds of motherboards, keyboards and mice. Anticipating the growing importance of research and development, the National Science Council of Taiwan published its first 'White Paper on Science and Technology' in December 1997. The paper outlined one major challenge for Taiwan on the eve of the twenty-first century as being 'The arrival of an "information society"'. This was said to be 'deeply affecting the way people live and work, changing the way companies and the government operate, and bringing about a new culture in a world that is fast becoming an even more closely-knit global village'.[12]

In order to meet the challenge of this emerging information society, the National Science Council recommended that Taiwan should develop into a 'technologically advanced nation' during the first decade of the twenty-first century. Deploying the image of Silicon Valley-type 'clustering', it envisioned the building of a 'National Information Infrastructure' (NII), starting with the establishment of a suitable 'core' and 'satellite science-based industrial parks/clusters' throughout Taiwan. These would serve as nuclei for building 'science cities', to be linked together by various major infrastructural networks that would turn Taiwan into a 'science island'.[13] This scientific-technological construction of Taiwan's economic future was translated into more concrete measures and a specific timetable in the form of an 'Action Plan for Building a Technologically Advanced Nation', which appeared in April 1998. In relation to the establishment of hi-tech clusters, this recommended the building of new satellite industrial parks at Chunan and Tungluo as well as software parks at Nankang and other locations.[14]

This vision of Taiwan as a 'science island' was reinforced and deepened by the coming to office of President Chen Shui-bian in May 2000. Chen set the tone for this when he made a speech discussing UK Prime Minister Tony Blair's ideology of the 'Third Way' during a visit to London in December 1999, when he was still the pro-independence Democratic Progressive Party's (DPP) presidential candidate. According to this, he identified what he called the 'New Middle Way' as Taiwan's future path in the context of 'globalization'. It was also in this speech that Chen proposed his vision of Taiwan as a 'Green Silicon Island'. To get the flavour of what Chen said, it is worth quoting the relevant passage in full:

> I am both optimistic and worried as I look to Taiwan's future economic development under globalization. I am worried and concerned about Taiwan's lack of natural resources. . . . At the same time, my cause of optimism arises from Taiwan's comparative advantage in information technology. The high-tech industry concentrated in Hsin-chu and surrounding areas has become the base for the world's computer hardware, and Taiwan's software development is acquiring global stature. This is Taiwan's chief opportunity.
>
> Over the years, I have had a vision of developing Taiwan into a Green Valley. I believe that human beings are entitled to enjoy a beautiful natural environment as well as the convenience of advanced technology; I cannot imagine an essential conflict between the two. . . . My blueprint for Green Valley must be extended to the entire island, based on the current successes and resources of Taiwan's silicon and computer high-tech industry. I hope that Taiwan in the next millennium will indeed become the Green Silicon Island.[15]

Chen was quick to return to this theme when he gave his inaugural speech as Taiwan's new President on 20 May 2000, in which he envisioned the development of a 'sustainable green silicon island' that provides a 'balance between ecological preservation and economic development'.[16] Since then, the idea has found more concrete expression in the 'Plan to Develop a Knowledge-based Economy in Taiwan', produced in September 2000. This claims to be 'one of the driving forces for Taiwan's transformation into a "Green Silicon Island"', a project that can be implemented in six ways:

1 Set up mechanisms to encourage innovation and foster new ventures.
2 Expand the use of information technology and the Internet in production as well as daily life.
3 Lay the groundwork for an environment supportive of Internet usage.

'Greater China' and siliconization   107

4   Consider due modification of the education system in a drive to meet the development of personnel needs by training and importing a sufficient pool of knowledge workers.
5   Establish service-oriented government.
6   Formulate precautionary measures against social problems that arise from the transformation of the economy.[17]

From 'crisis' to 'bubble'

It is the hi-tech crisis discourses described above as emerging at the official level in Mainland China, Hong Kong and Taiwan, along with their imagined IT and/or biotech futures, that are generating new techno-economic identities and practices in the 'Greater China' region. Interested actors freely deploy the symbols of 'Silicon Valley', 'Internet', 'information technology', 'biotechnology', and 'innovation' to market themselves as new objects of 'future growth/development' that go beyond low-cost production, represented by showpiece cases such as the 'multimedia hub' and 'Green Silicon Island'. It is thus that the official rhetoric of siliconization generates discursive spaces and new forms of knowledge which key corporate actors can use to mutually reinforce each other's technological 'optimism'.

Needless to say, some of the projects that have emerged as the new object(ive)s of 'future growth' and even as 'hi-tech' icons from this technological optimism have fallen out of favour since the bursting of the 'technology bubble' in May 2000. The Cyberport is one example, presented as it was as Hong Kong's 'strategic cluster',[18] seeking to capture global 'information flows' and to manage these within the service-space of Hong Kong and the broader region. Such a project strongly resonated with a global–regional–local epistemic community comprising local capital (for example, Richard Li of the Pacific Century CyberWorks), the Hong Kong Government (the Chief Executive, the Financial Secretary, the Secretary for Information and Broadcasting), quasi-governmental organizations (Hong Kong Industrial Technology Centre), and global–regional capitalists (such as Microsoft's Bill Gates, Yahoo!'s Jerry Yang, and IBM's Craig Barrett). The latter group of 'cyber-gods' even flew to Hong Kong in early 1999 to publicly endorse the project and highlight their role as future tenants, helping to create a global–local epistemic community consisting of a cyber-elite that can reinforce itself in and through the discourse of 'information technology'. This network was expanded still further when Tung Chee-Hwa paid a personal visit to the original Silicon Valley in July 1999.

Concurrent with the hi-tech spectacle of Hong Kong's Cyberport, several Chinese cities in the Pearl River Delta region also sought to ride the information technology wave after the 1997 financial crisis. In 1998, for example, the Guangzhou municipal government signed an agreement with Hong Kong to expand cooperation in the fostering of new and hi-tech businesses, including plans to enhance greater cooperation between their colleges and universities in this field. The idea emerged of building a new 'hi-tech' corridor along the Guangzhou–Kowloon railway, which would include the cities of Shenzhen, Dongguan, Huizhou, Zengcheng and Guangzhou. It was in pursuance of this goal that Shenzhen hosted the October 1999 hi-tech fair at which Zhu Rongji gave his key speech, considerably boosting China's hi-tech imagination and consolidating the city's 'hi-tech' identity.

The Shenzhen event was attended by actors at the global, regional and local levels of activity. Senior PRC officials included the Minister of Science and Technology, the Foreign Trade Minister and the Minister of Information Industry. Leading national firms like Huawei Technologies and Great Wall Computer were present, as were individual hi-tech icons like Chinese-American Steven Chu (winner of the 1997 Nobel prize for physics), Hong Kong Financial Secretary Donald Tsang, and former Japanese Prime Minister Kaifu Toshiki. Delegates from universities throughout 'Greater China' also appeared, along with big multinational players such as Microsoft, IBM and Epson. This group of organizations and individuals expressed a common hi-tech voice that helped to reinvent Shenzhen as a 'gateway of information technology' in and through their exhibits, speeches, lectures, and on-site visits.[19] Following Shenzhen's spectacle, the city of Dongguan soon followed with the first 'Computer Information Product Exposition', in October 1999. No less than 75 per cent of the exhibitors came from Taiwan, many of whom were based in the island's Hsinchu Science Park.

Fearful of being left out of the information race, Taiwan was in fact also busy developing its plans to promote hi-tech clusters besides Hsinchu. The original park's founders, such as scientific and economic guru Li Kuo-ting, were frequently deployed as icons to highlight the successful transfer of the 'Silicon Valley model' to Taiwan, with the island symbolized as a 'high-flying graduate' of the California experience, specializing in semiconductor OEM production. In the context of growing fears that software production was falling behind Taiwan's extremely successful hardware industry, the Executive Yuan's (Cabinet) 1998 'Action Plan' put the Silicon Valley metaphor to work in extending Hsinchu's 'success' to cover the whole of the island. The older vision of Taiwan as a 'science island' was then later reinforced by Chen Shui-bian's vision of a dynamic 'Green Silicon Island' established on knowledge-based technologies.

This new economic identity was promoted by Chen Shui-bian's administration and local government leaders like Taipei Mayor, Ma Ying-jeou, whose city hosted the 2000 World Congress on Information Technology. It has also been strengthened by prominent business figures like Stan Shih of the Acer Group, a community of well-connected Taiwanese-Americans, and personnel exchanges with California's Silicon Valley. The members of this trans-Pacific community are sometimes known as 'astronauts' because their work as engineers, executives and 'angel' investors (venture capitalists) in Taiwan and the United States means that they spend most of their time on aeroplanes. They have thus come to play a key role in coordinating Taiwan's financial and manufacturing strengths with Silicon Valley's engineering and research skills through their mobility and Internet networks.[20]

The Acer Group actually ran a number of Websites to link the community of 'astronauts' together. One of its well-known sites was a Chinese-language portal called the *Silicon Valley Journal*, which deployed the 'new economy' discourse and profiled itself as providing 'Chinese Wall Street Reports'. It played an important role in linking the Chinese-American community in 'the Valley' with the Asia-Pacific region in and through the 'Silicon Valley' dream. This dream was explicitly presented as 'a kind of metaphor of hi-tech culture' that symbolizes 'innovation', 'entrepreneurialism', 'networking', 'mobility', 'clustering', 'risk taking', and (at least before the 'technology bubble' burst) an 'unprecedented gold rush'. The publication of the 'success stories' of notable Internet 'whizzkids' was particularly helpful for encouraging the 'gold rush' image, with the wealth of Chinese-American 'infopreneurs' highlighting the nature of the 'Internet dream' (Yahoo!'s Jerry Yang was said to be worth close to USD 4.8 bn!).[21]

New techno-economic practices: assemblage and struggle

It has been described above how the celebration of the 'Silicon Valley' imagination came to strongly resonate with regional discourses of Taiwan as the 'Green Silicon Island', Hong Kong as an 'e-hub', and Shenzhen as 'China's Silicon Valley' after the Asian financial crisis. Although the bursting of the 'technology bubble' has decelerated the pace of informational development, such discourses concerning reinventing economies in 'Greater China' around 'hi-tech' accumulation have remained popular. Strategic actors still use the symbolism of 'hi-tech', 'B2B commerce', 'biotechnology', and 'Silicon Valley' in their continuing efforts to (re-)ground new techno-economic identities and practices.

Whether the drastic market adjustment that began in May 2000 will affect enthusiasm for 'hi-tech' and 'information technology' in the longer term remains to be seen. Nonetheless, the techno-economic knowledge that has already been produced is affecting the building of new practices across different sites, albeit at a slower and more cautious pace since the 'bubble' burst. These practices include the following.

- Developing flagship 'incubators' that profile themselves as 'the next Silicon Valley' and are promoted by private–public partnerships.
- Building new regional–global networks with Silicon Valley in California and with analogous clusters elsewhere (in Singapore, for example).
- Forming alliances between the 'old' and the 'new' economies and promoting their interpenetration.
- Re-articulating the global–regional–national–local scales of activities for hardware, software and Internet delivery.

- Developing new sources of networking (such as industry–university cooperation) and finances for hi-tech ventures.
- Tapping overseas/Chinese ICT experts through schemes like the 'Admission of Talent Scheme' in Hong Kong and new 'visa wars', in which countries in the region issue visas to compete for the limited pool of hi-tech workers.

In the rest of this chapter I will examine these practices and the way they are assembled into a geo-economic regime both within and across borders. It should be noted first of all, though, that such assemblage takes place not in accordance with the will of the government or individuals, but by the articulation and concurrent development of practices with diverse trajectories. The resulting contingent assemblages can be analyzed in terms of the co-presence of cooperation with competition and contestation in what amounts to a geo-economic regime.

Hong Kong

Let us start with the case of Hong Kong, which began to move its labour-intensive industries to southern China following the opening of the PRC in 1978. When the partial vacuum created by this 'hollowing out' process was filled by Hong Kong's acquisition of functions that made it into a global–regional gateway city, the result was a heavy dependence on the service sector – especially real estate and finance. A major debate on 'service versus industry' thus raged from 1993 onwards,[22] which became even more pressing when the Asian financial crisis demonstrated the vulnerabilities of the real estate and financial sectors in Hong Kong. It was in this context of the desperate post-financial crisis search for new object(s) of economic growth that Tung Chee-Hwa made his 1998 policy speech emphasizing 'hi-technology', using it as an economic symbol that might be capable of expanding the boundaries of the debate on Hong Kong's future trajectory. It was not long after, in March 1999, that his technological mode of calculation was taken up and articulated as the idea of building a Cyberport by Pacific Century's Richard Li.

Li's Cyberport is premised on the possibilities for expanding and upgrading the service sector. He thus describes it as providing 'a comprehensive facility designed to foster the development of Hong Kong's information services sector and to enhance Hong Kong's position as the premier information and telecommunications hub in Asia'.[23] Using 'Silicon Valley' and its 'social density' as metaphors, it is alleged that the Cyberport will be able to 'attract, nurture and retain the relevant innovative talent necessary to build a cyber-culture critical mass in Hong Kong'.[24] Within this narrative, the Cyberport is represented as a new type of service-based node for connecting Hong Kong to the fast-time of 'information flows'. More specifically, it is a project that re-imagines the territory's competitive advantages in terms of capturing global 'information flows' and managing them within the service-space of Hong

Kong and its broader regional scale through, for example, the formation of a multimedia and information services hub. In addition, Hong Kong's services are to be connected to fast cyber-time and the knowledge-based economy. Finally, a localized social space is to be consolidated within which to build a 'cyber culture critical mass' that links the global, the regional and the local to consolidate a 'pool of talents' in the shortest possible time.

This vision of expanding and upgrading Hong Kong's service cluster(s) has clearly been appropriated by the government and used to symbolize and spearhead its post-financial crisis politics of 'technological optimism'. It is thus that Financial Secretary Donald Tsang earmarked the 'Cyberport' as a flagship project in his 1999 budget speech, involving a private–public partnership in which the government provides land worth HKD 6 bn, while Pacific CyberWork pools capital to construct the buildings. This project is then supposed to provide 'incubator' services to create a 'critical mass of firms' that will be nurtured by a physical form modelled on 'Silicon Valley' (Table 5.1).

Not surprisingly, this emergent discourse and its private–public practice encounters resistance from within the service sector and from political groups. Seven property developers have publicly denounced the project, and been joined by three more, angry at being excluded from the high-profile project by the government's decision to provide free land for the Cyberport without any public tendering process. In response, the Financial Secretary argued that the Cyberport was a 'technology' and not a 'residential project', explaining:

> We have not tried to exclude anybody in the process. . . . The whole emphasis is Cyberport. The whole emphasis is technological project. It is not residential development as such. Even if you use all the money that we have from selling the residential portion, we will not have sufficient capital to develop the Cyberport. So the way we are approaching the issue is the most economic, most efficient way, from a taxpayer's point of view. . . . You must realize of course the Cyberport, at the end of the day, will be owned by the public, will not be owned by any private developer. We will be determining the leasing requirements. We will be determining who will be our tenants, and we will be determining the rental value as well. So for someone who has expertise, who has the connection, and is able to put up with the business risk for this matter, and is a technology firm, it is a very rare find. I think we have got the right thing, and we've got the right deal for the Hong Kong public.[25]

This definition of the Cyberport as a 'technology project' did not entirely pacify the discontented developers, who still saw it as a 'residential project'. They even came up with an alternative proposal, according to which the government could auction off the ancillary residential property and receive USD 1.08 bn in cash up front, of which USD 640 million could be used to construct the Cyberport. The government rejected this by appealing to the

Table 5.1 A new economic object of 'growth': Cyberport

Cost:	HKD 13 billion (USD 1.68 bn)
Size:	64 acres (25.6 hectares)
Location:	Telegraph Bay, Pokfulam
Aims:	'To create a world class location for the conduct of a variety of activities which through the use of information technologies, can leverage Hong Kong's existing strengths in the service sector (e.g. in financial, media, retail, transportation, education, and tourism services).'[a]
Built environment I:	Cyber facilities (2/3 of the site) • fibre optic wiring • satellite signal senders • built-in high speed modems • cyberlibrary • media laboratories and studio facilities
Built environment II:	Real estate (1/3 of the site) • houses and apartment • hotel • retail
Completion date:	2007 (commencing from 2002)
Job creation:	4,000 during construction 12,000 professional jobs on completion (10% from outside Hong Kong)
Partners:	Pacific Century CyberWorks (HK$ 7 billion equity capital). Government (land worth HK$ 6 billion).
Cluster:	Multinational corporations (Microsoft, IBM, Oracle, HP, Softbank, Yahoo!, Hua Wei, Sybase). Local tenants of small to medium-size information technology companies.
Metaphors/images used:	'Silicon Valley', 'catching up', and 'clustering'.

Source: Various issues of *South China Morning Post*.
Note
a Hong Kong Cyber-Port, 'What is Cyber-Port?', p. 1.

commercial logic of 'risk calculation', according to which the private–public partnership would mean less 'risk' for the government.

It was not only property developers who saw the Cyberport as constituting a 'residential' project. Some financial market analysts also criticized the project for amounting to little more than 'Cyber villas by sea', claiming that it was 'no "Silicon Valley"'.[26] The Hong Kong Democratic Foundation, moreover, adopted an even harsher tone, responding to the 1999/2000 Budget by remarking that:

> We do not believe that a property-based development is a meaningful way to promote high technology industrial development. The clustering

of technology-related industry does not depend on property; there are already small scale clusters in areas of Hong Kong that have received no special favour, for example, the Wellington Street area.[27]

The Democratic Foundation's report also criticized the granting of the Cyberport land to a well-connected company without any tendering procedures for raising the spectre of cronyism and damaging Hong Kong's reputation. Members of the Democratic Party echoed this charge, and challenged the government's lack of transparency, pointing out that Richard Li is the son of Li Ka-Shing, a good friend of the Chief Executive. The Financial Secretary responded as follows:

> I do not want to comment on what other people say, but I will be very patient in explaining to them this is a very important infrastructure, and we have been doing it very fairly. But the requirements for selection of a developer are very strict because it is not a property development project. And I would be very careful in explaining cronyism is never, never in Hong Kong's dictionary. We pride on being transparent, we pride on playing on a completely level playing field. There is not [sic] question whatsoever of Hong Kong Government, the SAR Government, engaging in cronyism.[28]

Despite these challenges to the government's plan for a private–public partnership, the emerging techno-economic discourse has generated other new economic practices within the service sector. Until the bursting of the technology bubble, one notable case was the way in which large property and commercial conglomerates in Hong Kong began to combine the so-called 'old' and 'new' economies with varying degrees of success. For example, in August 1999, Sun Hung Kai Properties (SHKP) transformed its empty properties in Tsuen Wan to establish a 'Cyberincubator' project in partnership with the Hong Kong Industrial Technology Centre. Under this scheme, SHKP would provide rent-free space for new 'infopreneurs' for three years in return for 15 per cent stakes in their respective businesses. The response to this initiative was poor, however, due to the terms involved, with start-up firms believing that the value of their equity stake was much higher than the rental that could be saved from receiving the free industrial space being offered by SHKP. In May 2001, the Hong Kong Industrial Technology Centre abandoned this rent-for-equity model.

Apart from this kind of experimentation to find new uses for its old property assets, SHKP also stretched into the 'new economy' by developing its Internet and China-related projects. One of these was the 'SUNeVision' plan to promote its information technology infrastructure and Internet services, such as online selling of property and insurance. Another was a USD 20 million 'C Tech Fund' with SmarTone to focus on venture capital investments in IT, healthcare and biotechnology, the environment, telecommunications and other technology-related fields in Mainland China. Needless to say, SHKP's attempts to stretch into the 'new economy' have slowed down considerably since May 2000.

Tung Chee-Hwa's July 1999 visit to Silicon Valley also fed into the creation of new practices by building new linkages between local informational communities and the Chinese diaspora. The Hong Kong–Silicon Valley Association and a new Website (SV-Hong Kong.com) were formed in late 1999 to enhance possible global–local flows of knowledge, expertise and labour. The Hong Kong Stock Exchange also began to experiment with new initiatives, such as the launching of the Growth Enterprise Market (GEM)[29] in 1999, the territory's version of Nasdaq, offering an alternative listing choice for incubating start-up technology companies and raising venture capital in the Greater China region. Up to the bursting of the bubble, 13 firms were listed, with 7 from Hong Kong, 4 from China and 1 from Taiwan. One of the high-profiled listings was Tom.com, an Internet arm of Li Ka-Shing's Cheung Kong-Hutchinson Whampoa empire. Li, identified as the 'superman' of property, telecommunications and port facilities, managed to reignite the dizzying pre-crisis speculation craze. Five hundred thousand investors mobbed local branches of the Hongkong and Shanghai Banking Corporation to deposit applications for Tom.com's initial public offering (IPO) in February 2000. The IPO was oversubscribed 2,000 times due to an 'Internet fever' that was largely related to people seeking to earn 'a quick buck' in the same way that they had once over-invested in the property market. In this regard, the 'Cyberport', 'Silicon Valley', 'Richard Li', and 'Tom.com' became short-term economic icons just before the bursting of the 'technology bubble'.

When the bubble did finally burst, it not only brought falling share prices and retrenchment in high-profile corporations, it also revealed the conflictual nature of the assemblages of techno-economic practices that had developed in Hong Kong. First of all, the use of the Cyberport as a symbol to trigger the expansion of the 'old' economy into the 'new' failed to bring about the intended expansion of existing economic boundaries to embrace 'technology'. Instead, it became evident that it had spurred Hong Kong's embedded property-finance interests to react in a short-term euphoric manner. Second, the short-term 'hi-tech' fanfare, if not 'spec-fare' (meaning to do well from speculation), ignited by the service-finance sectors and the subsequent bursting of the technology bubble actually stimulated a good deal of public adversity towards 'technology'. In a paradoxical way, therefore, a project aimed at promoting 'technology' ended up doing the opposite. 'Technology', largely understood in speculative terms, is neither facilitating Hong Kong's catching-up process nor bridging the service-industry divide.

Finally, global competition to capture the fast-time of informational capitalism in Hong Kong has changed the temporalities of its decision-making process, most notably by speeding up the building of the Cyberport. Such speeding up, however, came into direct conflict with the routine procedures of the public tendering process upon which business confidence depends. With

Taiwan

these problems in mind, Hong Kong's post-financial crisis regime is now searching for new objects around which to reconstruct its mechanisms of governance. At the time of writing, the ideas of the 'world city' and the 'logistic centre' are gaining new prominence.[30]

Taiwan

The case of Taiwan has developed somewhat differently from that of Hong Kong due to the advantageous position of its electronics industries since the 1980s. Although the earthquake that struck the island in September 1999 caused some setbacks, no long-term damage was inflicted on its production capacity. Hsinchu Science Park has remained a manufacturing powerhouse with good connections to California's Silicon Valley and the mantra has been 'Silicon Valley creates it, Taiwan makes it'. In accordance with the 1997 White Paper and its Action Plan, Taiwan is upgrading its research capacity by developing other clusters in Tainan, Chunan and Tunglu.

Apart from building industrial parks, in October 1999 then President Lee Teng-hui inaugurated the opening of the 'milestone-type' Nankang Software Park in Taipei county (Table 5.2). This flagship public–private project was commissioned by the Ministry of Economic Affairs and developed by Century Development Company, a joint venture of 19 domestic and foreign companies. The International Software Development Centre was established on 2,000 ping (1 ping = 11 sq. metres) of floor space purchased by the Ministry, and operates to encourage the grounding of global–local links between multinational software companies and indigenous Taiwanese firms. In addition, the Ministry will spend NTD 995 million to procure 3,100 ping of floor space to set up an incubator programme for around 60 start-up

Table 5.2 Nankang Software Park

Cost:	NTD 12.8 billion (USD402.5 million)
Size:	8.2 hectares
Location:	Nankang, Taipei County
Completion date:	1999 (first stage); 2003 (second stage)
Partners:	Century Development Corp.
Cluster:	Home to global and local software companies. Proximity to the Academia Sinica's Institute of Information Sciences and the projected Nankang Economic and Trade Park.
Signed-up tenants:	15 (including IBM, HP, Compaq, Intel)
Terms:	Foreign companies have to sign cooperative agreements with Taiwan companies.
Expected return:	USD 14 bn by 2005

Source: Various issues of *Central News Agency* reports.

companies. The park thus aims to provide software companies with infrastructure facilities, networking services, training programmes, and market information. Software companies that locate there may also take advantage of tax incentives and subsidies for developing strategic technology.

Riding the silicon wave, hardware producers in Taiwan are also rethinking their strategies, with new practices emerging. The way in which hardware conglomerates are diversifying into the 'new economy' can be seen in the way that the Acer Group formed the Acer Digital Services Corporation in 1999. This has embarked on a number of activities, including establishing a global–regional alliance with Cisco Systems and General Electric Information Services. This operates an Internet shopping mall under the AcerNet, developing X-media for media display to be used in the convenience chain store under Web Point and developing and marketing children's software services. It has also entered the venture capital business by funding start-ups in the United States and the Asia-Pacific region under Acer SoftCapital Group.

From a political perspective, this kind of diversification into the 'new economy' is less controversial than the practice of relocating production to cheaper sites in Mainland China, a process that began with low-tech products such as umbrellas and footwear in 1987. The present round involves more advanced products, such as keyboards, mouse technology and switching power supplies.[31] This trend of relocation to the Mainland has become increasingly controversial as it has coincided with the decline of export demand in Taiwan and the decline in political confidence that has accompanied the political transition from Kuomintang to DPP rule.

Given that Taiwan has a longer history than Hong Kong of pursuing the strategy of siliconization, it has consolidated a 'Silicon coalition' of powerful industrial capitalists, exemplified by figures like Acer's Stan Shih and Matthew Miao of the Mitac-Synnex Group. Taking advantage of China's cheaper production costs[32] and its large potential markets, this group has expanded production in Mainland China in the same way as industrial capitalists such as Wang Yung-ching of Formosa Plastics Group have. For example, the Acer Group invested USD 50 million in late 1999 in a manufacturing plant in Zhongshan, Guangdong Province, to produce computers and DVD players. In 2001, the company produced about 122,000 mainboards per month, 250,000 CD-ROM drives and 50,000 bare-bones systems in its Zhongshan plant. It has been estimated that 30 per cent of Taiwan's total IT production was made in Mainland China in 2000.[33]

This extension of Taiwan's Silicon chain 'westward' has sparked new power struggles in Taiwan. In order to add credibility to its cross-Strait activities, the 'Silicon coalition', with the support of other capitalists and the political opposition alliance, deploys a 'China-as-partner' discourse that narrates 'China' as an 'external economic boost for Taiwan'. By mapping Taiwan's economic future with that of China, this 'Silicon (plus) coalition' calls for the Taiwan government to relax its 'no haste, be patient' policy[34] towards the Mainland by lifting all curbs on investment across the Strait. Wang Yung-ching of Formosa Plastics even suggests that Taiwan should accept Beijing's 'one-China' principle. The 'win–win' discourse of this 'Silicon (plus) coalition' is also articulated by a plethora of commercial books that mediate the negotiation of China's identity in Taiwan. Deploying titles such as *The Winning Commercial Potential of 1.3 Billion Chinese, Dipping into the Golden Bowl of Shanghai, and Thirty-Five Gold-Panning Measures for the Mainland Stock Market*, such works construct an image of Mainland China as a 'gold mine' and Shanghai as a 'cool city' in which Taiwanese capitalists and professional people can 'live, work and play'.[35]

Such positive constructions of China coexist with the new political and economic practices that are promoted by the 'Silicon (plus) coalition'. For example, pressure is put on the government to abandon restrictions on hi-tech industries, such as notebook computer production and semiconductor manufacturing. Firms also by-pass government policy and invest in notebook computer production in China through subsidiaries and holding companies in third places (such as Hong Kong), something that nine out of ten notebook makers have already done. They also enter into the Chinese Internet, telecommunications and hi-tech markets. Acer Inc., the Internet company established under the Acer Group, is even considering going public in Mainland China rather than on the Nasdaq.

This 'go-west' imagination and its associated practices have raised considerable political concerns in Taiwan. Vice-President Annette Lu has dubbed it the 'Mainland Fever', worrying in an economic forum how, 'This fantasy about the mainland has triggered uncertainty about the future of Taiwan, causing the relocation of capital and production bases. . . . Blindly following the mainland fever without knowing the risk behind it is not something a wise man does'.[36]

In subsequent speeches, Lu has repeated the same economic worries and urged the Taiwanese business community to keep its roots in Taiwan. Apart from economic concerns, it was also reported by the *Hong Kong iMail* that Lu described Mainland China as the 'People's Republic of Cheats' in a meeting with a group of Taiwanese investing in parts of Asia, excluding China.[37] This public display of conflict has reignited the 'hollowing out' discussion in Taiwan and the geo-political and geo-economic dilemmas that the island has to face. More specifically, there are increasing fears over rising unemployment and the outflows of capital and labour to China, with some 300,000 personnel, mostly managers, reported to have moved across the Taiwan Strait.[38] Such trends sharpen Taiwan's dilemmas, underlying which there is not only the fear of over-dependence on Mainland China, but also the international and domestic implications of the possibility that Taiwan's position as a major global supplier of computer hardware will be overtaken. Annette Lu summed up the paradox of Taiwan's position when she remarked that it is 'ridiculous' for Taiwanese companies to pour money into China, when Beijing has 300 missiles aimed at the island.[39] Under pressure to cut costs in the global supply chain, however, Stan Shih of Acer has remarked that the 'westward' march is 'inevitable'.[40]

In November 2001, after consultation with the Economic Advisory Council, President Chen Shui-bian finally agreed to loosen his predecessor's 'no haste, be patient' policy, raising the ceiling on Mainland investment by Taiwanese companies to USD 80 million. In order to strike the right balance between geo-economic and geo-political demands, officials of the DPP have suggested a return to 'the spirit of 1992', referring to a consensus that the two sides of the Strait would accept that each had its own interpretation of the meaning of 'one China'.[41] The hope that this could revive friendly relations with the PRC without conceding to its pressures for accepting 'unification', however, has been dashed by Beijing's insistence on the priority of its nationalist geo-political concerns on the issue of Taiwan's status. With the DPP having consolidated its position as the ruling party in Taiwan in parliamentary elections held in December 2001, the 'One China Principle' has continued to be a bone of contention between Taipei and Beijing. While the DPP is stronger at home, it is left facing the challenge of hard security concerns that arise from growing economic integration between the two sides of the Strait.

Southern China

The area of the Pearl River Delta and Guangdong Province, bordering Hong Kong, has seen numerous attempts to build hi-tech industrial development zones in the cities of Shenzhen, Zhuhai, Huizhou, Zhongshan and Foshan, under the auspices of central government organs like the Ministry of Science and Technology, as well as municipal authorities. Most of these projects include the offer of tax exemptions and/or the opportunities of domestic sales for multinational and national firms who enter the zones. The Guangzhou Economic and Technological Development District was one of the first state-level development districts, and merged with the Guangzhou Hi-Tech Industrial Development Zone in 1998. One of its recent flagship projects has been to develop the Guangzhou Science City, which profiles itself as 'the Rising Silicon Valley in Guangzhou'. As a state-run project, it offers tax exemptions for investors for the initial two years and unlimited sales in Guangdong for hi-tech electronics, computer communication and aerospace engineering products.

Shenzhen's Hi-Tech Industrial Park in the Pearl River Delta area has been successful in attracting a host of multinational and national technology firms, such as IBM and China's own Great Wall. In 1999 these firms produced electronic and ICT products – such as computer component parts, mobile telecommunication products and computer software – with an output value of RMB 29.7 bn. Apart from this state-run park, the Shenzhen Municipal Government is also encouraging enterprises to develop research and development centres and private–public partnerships. In 1999, there were 271 research and development centres in the city, some of them having connections with Silicon Valley in California as well as with similar set-ups in Beijing, Shanghai and Nanjing. The central government is also encouraging national and overseas universities to cooperate with local concerns. For example, Beijing's Qinghua University was granted 10,000 square metres of land to build a research and development centre and RMB 60 million for research projects by the Shenzhen Municipal Government. Beijing University, along with the Hong Kong University of Science and Technology and the Shenzhen Municipal Government, have together invested RMB 80 million in an education and research centre. This public–private partnership runs an MBA programme, conducts biotech research and is home to iSilk.com, a firm that is developing simultaneous translation software for the Internet.[42]

Private–public partnerships have been further developed in a new flagship project approved by the Ministry of Science and Technology and the Shenzhen Municipal Government. This involves the establishment of the Mainland's first venture capital fund to build a hi-tech park – CyberCity Shenzhen (see Table 5.3). The scheme was spearheaded by Simon Jiang (the son of former National People's Congress chairman Qiao Shi), in conjunction with Hong Kong investment company Cyber City Holdings and United States venture capital company Dynamic Technology Inc. Based on the symbolism associated with 'Silicon Valley's high-risk, high-return promise',[43] its aim is to nurture and prepare software and Internet firms for the market by connecting them with overseas firms and investors.

Since its inception, this experimental private–public partnership has been extended in two spatial directions. First, it has penetrated the Hong Kong financial sector by arranging a backdoor listing on the Hong Kong Stock Exchange by selling 80 per cent of its stake to a China-owned but Hong Kong-listed company, Hing Kong Holdings. Then, in April 2000, it extended its network to the 'Intelligent Island' of Singapore by selling a total of 23.3 per

Table 5.3 CyberCity Shenzhen

Venture capital:	USD 250 million private equity fund (Memorandum of Understanding regarding USD 100 million with DynaFund Venture).
Land:	80% below market price from Shenzhen Municipal Government.
Size:	36,000 square metres (40.5 hectares)
Completion date:	2004
Partners:	Hong Kong-based CyberCity International
Cluster:	Home to 100 international and local software and information companies.
Signed-up tenants:	60 global and local firms.
Listing:	Selling 80% stake to Hing Kong (a company listed on the Hong Kong Stock Exchange).
Network:	Cooperation agreement with 10 mainland software parks (e.g. Kunming, Tianjin, Pudong, Xian).

Source: Various issues of *South China Morning Post*.

cent of its stake to Temasek Holding (an investment arm of the Singapore government) and to Fraser and Neave (a Singapore consortium). A new company called Vision Century Ltd was formed under the control of Singapore capital.[44] Such a development is particularly interesting when Singapore is itself relinquishing its control of the Suzhou Industrial Park – that it initially financed – to its Chinese partner. This refocusing of Singapore's interest on southern China thus indicates the continuing possibility of a synergy of 'software transfer' and transborder practices stretching beyond the original 'Greater China' region of the Mainland, Hong Kong and Taiwan.

Cooperation or competition

The preceding discussion of new material practices in 'Greater China' helps to resolve the question of whether the strategy of siliconization will help to strengthen the competitiveness of the region. In respect of Taiwan, it can be argued that the relocation of industries to Mainland China and the upgrading of its own technological base enables the island to concentrate more on research and development, the global connections of its computer and IT companies and possibly on the supply of Internet content as well. In the case of Hong Kong, its role as a traditional entrepot city is declining with the emergence of other ports in the Pearl River Delta, and the Asian 'crisis' has also exposed an over-dependence on property and financial markets. The territory is thus repositioning itself as a 'logistic, financial and digital centre' with hard and soft infrastructure for Internet services (broadband networks, e-commerce, business consultancy, data centres, content distribution, marketing skills, venture capital, GEM) and project finance. Given that Taiwan is relaxing restrictions on investments in Mainland China, where the demand for investment and venture capital is only likely to increase with WTO accession, Hong Kong is strategically positioned to act as a gateway and fund-raising centre for the whole 'Greater China' region. Likewise, Shenzhen and the Pearl River Delta are well placed to continue their concentration on electronics and IT products, as well as offshore software sites for Chinese-language and multilingual products.

Despite these possibilities for stronger cooperation and the rebuilding of cross-border governance mechanisms along the Silicon and informational supply chain in 'Greater China', however, competition also exists among the various incubators that are being developed. As shown above, the number of software parks in the region is growing, and their similarity may well lead to mutual competition. This can be seen in a number of areas, such as the provision of facilities to house software companies specializing in, for example, Chinese-language applications, or the incubation of local small start-up firms for joint ventures and market listings. The same can be said of the shared aim of attracting global players such as Microsoft, IBM and Oracle to relocate,[45] in the hope that other overseas and local technology firms will be drawn along in their wake. There will also be intense competition to lure talented personnel from other provinces within China and from among the overseas Chinese student community to work in the parks or elsewhere in the economy. As Shenzhen mayor Li Zibin put it in a May 2000 interview in *Asian Affairs*, '[Hong Kong] needs to attract talents from overseas and it is targeting the same sort of people as Shenzhen'.[47]

Apart from warning about the competition for talent, Li also remarked on the poor coordination between Hong Kong and Shenzhen when it comes to 'hi-tech' development, noting:

> In the high-tech industry, I have personally appealed two or three years ago for both sides to develop Research & Development on a common basis. It is really important to put our strength together and work for a common human resources development scheme. But it is not the case. In the financial sector as well. The two cities can work closer, but Shenzhen is much more willing to move on this topic too.

Even more significant for the emerging power dispensation within 'Greater China' is the way that Li Zibin then went on to interpret the Hong Kong–Shenzhen relationship in terms of a 'colonial' hierarchy, in which Hong Kong is seen to be better than Shenzhen and does not need Shenzhen, while Shenzhen needs Hong Kong, explaining:

> Some people at some level who deal with these questions in Hong Kong do not adhere to the view that to develop a closer relationship with Shenzhen and harmonize the build-up of high value industries will be beneficial to both. They look down on Shenzhen, and although we repeatedly emphasized the opportunities for each side to benefit from each other, they really believe that they will do well on their own. In this way, they are just keeping an old colonial mentality. They still ignore the fact that Shenzhen can develop without Hong Kong.[47]

Whether or not Li Zibin's interpretation of the relationship between Hong Kong and Shenzhen is correct, if such views are widely held then they could weaken the possibilities for inter-urban cooperation in southern China by encouraging a search for alternative partners. That would only further intensify competition in the 'Greater China' region.

Concluding remarks

This chapter has examined ways in which the idea of 'Silicon Valley' allows private and public actors to deploy the symbolism of 'high-technology' and the 'information age' to redefine the economic future within 'Greater China', especially since the Asian financial crisis. The strategy of 'siliconization' that this amounts to involves actors constructing and seeking to develop modes of coordination and governance around new regimes of hi-tech accumulation. This can be seen in a number of ways, such as the emergence in the region of several 'Silicon Valleys' to provide IT and incubation facilities for global firms and local start-ups. It is also evident in the changing power relationships between hi-tech concentrations, such as the emerging status of Shenzhen as a kind of 'junior partner' to Hong Kong. As for Hong Kong itself, the Cyberport project is repositioning the territory as a 'logistic, financial and digital centre', while Taiwan's information industries are being simultaneously upgraded and carried 'westward' to the Mainland by its 'Silicon (plus) coalition'. Among the other emergent phenomena worth mentioning is the introduction of the Growth Enterprise Market (GEM) in Hong Kong as a fund-raising avenue for hi-tech start-ups, the extension of diasporic silicon networking to sites in Asia and California, attempts by leading 'old economy' firms in the region to form strategic alliances or joint ventures in the hope of turning 'Greater China' into an e-production and e-service space, and the emergence of public–private partnerships to build the 'next Silicon Valley'.

On the increasingly inseparable regional, national and local levels of activity, intensified competition may create a deepening division of labour. For example, technological upgrading in Taiwan may lead to a concentration on research and development at home, while the manufacture of sophisticated components will take place in Mainland China. Hong Kong will become a 'logistic, financial and digital centre' with hard and soft infrastructure for Internet services and project finance, while Shenzhen continues to focus on electronics and IT products, as well as offshore software sites for Chinese-language and multilingual products. Yet if there is also competition between different 'Silicon Valley' projects, the development of a cross-border division of labour will proceed by way of both cooperation and contestation. It is too soon to predict whether this complex field of sometimes competing, sometimes complementary, practices pursued by private–public actors on different scales will produce a coherent mode of growth and governance in the longer term.

Of broader significance for regional politics is the way that the strategy of siliconization may impact on the process of nation-building in the different parts of 'Greater China'. The nature of this relationship is clearest where national identity is most contested, namely the relationship between Taiwan and Mainland China. Here it can be seen on the one hand that a transnational community of Taiwanese-Americans is emerging, enjoying close connections and personnel exchanges with California's Silicon Valley. On the other hand, the march of Taiwanese hi-tech investment to southern China has extended the Silicon chain westward, with serious implications for the island's domestic politics of identity.

One way to understand these emergent global–regional linkages between Silicon Valley in California and the information technology hardware clusters in Greater China, is in terms of a kind of 'Silicon Bridge' that is mediated by the cooperative efforts of private and public actors from all the territories concerned. Such agents form part of the backbone of the global hardware industry in which Taiwan and China are important nodal points in providing silicon-based products, such as computers and networking systems. Taiwan

'Greater China' and siliconization 123

supplied 53 per cent of the world's laptops and 25 per cent of its desktops in the year 2000. Thus, while growing integration between the various parts of 'Greater China' may act as a centripetal political force, it should be remembered that serious security crises across the Taiwan Strait would have profound effects upon the world's supply of IT products. This would, of course, have a knock-on effect on the development of informational capitalism in the United States, Japan and other developed countries. In this regard, the strategy of siliconization and the resulting 'Silicon Bridge' reflects both the deeper geo-economic potentials and the long-term strategic dilemmas that are involved in the remaking of the 'Greater China' region itself.[48]

Notes

1 On the regional division of labour in Greater China, see N. Sum, 'Rethinking Globalization: Re-articulating the Spatial Scale and Temporal Horizons of Trans-Border Spaces', in K. Olds, P. Dicken, F. Kelly, L. Kong and H. Yeung (eds), *Globalization and the Asia-Pacific: Contested Territories*, London: Routledge, 1999, pp. 129–46.
2 On the idea of natural economic territories, see A. Jordan and J. Khanna, 'Economic Interdependence and Challenges to the Nation-State: the Emergence of Natural Economic Territories in the Asia-Pacific', *Journal of International Affairs*, 1995, vol. 48, no. 2, pp. 433–62. The idea of the 'growth triangle' originates from Heng Toh-mun and L. Low (eds), *Regional Cooperation and Growth Triangles in ASEAN*, Singapore: Times Academic Press, 1995; see also Thant Myo, Tang Min and H. Kakazu (eds), *Growth Triangles in Asia: A New Approach to Regional Economic Cooperation*, Hong Kong: Oxford University Press, 1998. For a discussion on cross-border regions in Europe, Asia and Africa, see M. Perkmann and N. Sum (eds), *Globalization, Regionalization and Cross-Border Regions*, Basingstoke: Palgrave, 2002.
3 For further theoretical discussion of alternatives to the anarchy of market forces and hierarchical coordination by the state, see N. Sum, 'Time–Space Embeddedness and Geo-Governance of Cross-Border Regional Modes of Growth: their Nature and Dynamics in East Asian Cases', in A. Amin and J. Hausner (eds), *Beyond Market and Hierarchy*, Cheltenham: Edward Elgar, 1997, pp. 159–95. On an application of this approach to Greater China, see N. Sum, 'Rethinking Globalization', pp. 134–45.
4 For a pioneering study of Silicon Valley see A. Saxenian, *Regional Advantage: Culture and Competition in Silicon Valley and Route128*, Cambridge, MA: Harvard University Press, 1994.
5 On Silicon Valley as a habitat, see Lee Chong-moon, W. Miller, M. Hancock and H. Rowen (eds), *The Silicon Valley Edge: A Habitat for Innovation and Entrepreneurship*, Stanford, CA: Stanford University Press, 2000.
6 On Silicon Valley as a model of development, see D. Rosenberg, *Cloning Silicon Valley*, London: Pearson Education, 2002.
7 Quoted in Willy Lam Wo-lap, 'Hi-Tech Confucian Future' in 'Analysis', *South China Morning Post*, 13 October 1999, p. 8.
8 See *ChinaOnline*, 'Shenzhen Plans to Become China's Silicon Valley', 3 April 2000. Online. Available HTTP: <http://www.chinaonline.com/industry/infotech/NewsArchive/cs-protected/2000/april/B200032411.asp> (accessed 4 April 2001).
9 Tung Chee-Hwa, 'From Adversity to Opportunity', policy speech delivered by the Chief Executive in the Legislative Council Meeting, Hong Kong: Hong Kong SAR Government, 7 October 1998, p. 8.

10 Donald Tsang Yam-keung, 'Onward with New Strength', budget speech delivered by the Financial Secretary in the Legislative Council Meeting, Hong Kong: Hong Kong SAR Government, 3 March 1999, p. 15.
11 On the Chinese and Taiwanese professional associations in Silicon Valley, see A. Saxenian, 'Networks of Immigrant Entrepreneurs' in Lee Chong-moon, W. Miller, M. Hancock and H. Rowen (eds), *The Silicon Valley Edge*, pp. 248–75.
12 Executive Yuan, National Science Council, 'White Paper on Science and Technology', Taipei: National Science Council, Republic of China, 2 December 1997, p. 2. Online. Available HTTP: <http://www.stic.gov.tw/policy/scimeeting/E-whitepaper/summary_e.html> (accessed 27 July 2001).
13 Executive Yuan, National Science Council, 'White Paper on Science and Technology', pp. 5–6.
14 Executive Yuan, National Science Council, 'Action Plan for Building a Technologically Advanced Nation', Taipei: National Science Council, Republic of China, April 1998, p. 10.
15 Chen Shui-bian, 'The Third Way for Taiwan: A New Political Perspective', 6 December 1999. Online. Available HTTP: <http://www.president.gov.tw/1_president/e_subject-04a.html> (accessed 27 January 2000).
16 See Chen Shui-bian, 'President Chen Shui-bian's Inauguration Speech', 20 May 2000, p. 4. Online. Available HTTP: <http:// members.tripod.com/Ken_Davies/inaugural.html> (accessed 27 February 2002).
17 Executive Yuan, Council for Economic Planning and Development, 'Plan to Develop Knowledge-based Economy in Taiwan', Taipei: Council for Economic Planning and Development, Republic of China, September 2000, p. 8.
18 On the Cyberport as a strategic cluster, see Hong Kong Government, 'Cyberport: The Project', 15 April 2002. Online. Available HTTP: <http://www.info.gov.hk/itbb/english/cyberport/project.html> (accessed 27 May 2002).
19 For some of these details, see <http://home.ie.cuhk.hk/xtang/Introd.htm> (accessed 27 February 2002).
20 Hsu Jinn-yuh and A. Saxenian, 'The Limits of Guanxi Capitalism: Transnational Collaboration between Taiwan and the USA', *Environment and Planning*, 2000, vol. 32, no. 11, pp. 1991–2005.
21 The *Silicon Valley Journal* Website (http://www.syjournal.com/bingif/) changed its front page after the bursting of the technology bubble in May 2000. The discourses related to the Silicon Valley dream were replaced by more mundane information on new technologies. The Website had completely disappeared when the author attempted to re-access it on 27 February 2002.
22 On the industry versus service debate in Hong Kong, see B. Jessop and N. Sum, 'An Entrepreneurial City in Action: Hong Kong's Emerging Strategies in and for (Inter-)Urban Competition', *Urban Studies* (Special Issue on Asia's Global Cities), 2000, vol. 33, no. 3, pp. 2287–313.
23 See Hong Kong Cyber-Port, 'What is Cyber-Port?', p. 1. Online. Available HTTP: <http://www.cyber-port.com/whatis.html> (accessed 9 June 1999). Information on the Hong Kong Cyber-Port was no longer available when the author sought to re-access this site on 18 April 2002, but a copy is available from the author. However, the domain name does still exist and is up for sale. New information on the Cyber-Port is now posted on a new site. Online. Available HTTP: <http://www.cyberport-management.com/> (accessed 18 April 2002).
24 See Hong Kong Cyber-Port, 'What is Cyber-Port?', p. 1.
25 Donald Tsang Yam-keung, 'Financial Secretary's Transcript on Cyberport', Hong Kong: Hong Kong SAR Government, 17 March 1999. Online. Available HTTP: <http://www.info.gov.hk/gia/general/199903/17/0317146.htm> (accessed 6 December 1999).

26 David Webb, 'Cyber Villas by Sea', 22 March 1999. Online. Available HTTP: <http://www.webb-site.com/articles/cybervillas.htm> (accessed 12 December 1999).
27 Hong Kong Democratic Foundation, 'Policy Paper: Response to 1999/2000 Budget', 5 December 1999. Online. Available HTTP: <http://www.hkdf.org/papers/990512budget.htm> (accessed 6 December 2000).
28 Donald Tsang Yam-keung, 'Financial Secretary's Transcript on Cyberport'.
29 The Growth Enterprise Market (GEM) began operating on 25 October 1999 in Hong Kong. It was created to develop Hong Kong's information technology industry and is one of several Asian attempts to emulate the Nasdaq. The GEM is expected to serve the Greater China market, whereas Singapore's Sesdaq and Malaysia's Mesdaq are to serve the south Asian markets, while Kosdaq serves the Korean market. Some competition could come from China, where Shanghai and Shenzhen are reported to want second boards to compete with Hong Kong and the Nasdaq. However, Beijing currently prefers mainland non-state enterprises to seek flotation in Shenzhen and the GEM. The biggest competitor is the Nasdaq, which has attracted a number of initial public offerings from the region. Many venture capitalists are far more comfortable with the 28-year-old Nasdaq, given its liquidity and stable regulatory environment. Fearful of losing its edge, the Hong Kong Stock Exchange has eased its requirement for a lock-in period during which management are unable to sell their shares, from two years to six months.
30 On the new symbolism of the world city see N. Sum, 'Globalization and Hong Kong's Entrepreneurial City Strategies: Contested Visions and the Remaking of City Governance in (Post-)Crisis Hong Kong', in J. Logan (ed.), *The New Chinese City: Globalization and Market Reform*, Oxford: Blackwell, 2002, pp. 74–91.
31 N. Sum, 'Rethinking Globalization', p. 140.
32 In Mainland China, a computer engineer costs a quarter of one in Taiwan.
33 On Acer's investment in Zhongshan, see D. Baldwin, 'Acer Opens Biggest Mainboard Plant in Zhongshan', Nikkei Electronics Asia, April 2000 Issue. Online. Available HTTP: <http://www.nikkeibp.asiabiztech.com/nea/200004/cocn_98652.html> (accessed 29 March 2002).
34 Up to 2001, the Ministry of Economic Affairs and Securities and Futures Commission restricted China-bound investment to 40 per cent of a company's capitalization and 20 per cent of its net worth. The government limits individual investments to NTD 50 million.
35 For a discussion of this Taiwanese literature, see Tsai, Ting-I, 'Media Ignore China Investment Risks', 23 July 2001. Online. Available HTTP: <http://www.taipeitimes.com/news/2001/07/23/story/0000095302> (accessed 22 December 2001).
36 For a report regarding Lu's remarks, see *Taipei Times* Staff Writer, 'Lu Warns about Fantasy of China', *Taipei Times*, 19 January 2001. Online. Available HTTP: <http://www.taipeitimes.com/2001/01/19/print/0000070379> (accessed 23 May 2001).
37 On the *Hong Kong iMail* report, see Reuters, 'Taipei Blames Stagnant Economy on Cross-Strait Magnet Effect', *Hong Kong iMail*, 25 July 2001. Online. Available HTTP: <http://hk-imail.singtao.com/txtarticle_v.cfm?articleid=26215&intcatid=10> (accessed 25 July 2001). In the same report, *Hong Kong iMail* also highlighted that the reference to China as the 'Peoples Republic of Cheats' was missing in a statement issued by the Presidential Office. The term 'Peoples Republic of Cheats' appeared originally in *Far Eastern Economic Review*, 21 June 2001.

38 On the outflow of Taiwanese labour to China, see M. Forney, 'Taipei's Tech-Talent Exodus', 21 May 2001. Online. Available HTTP: <http://www.time.com/time/asia/news/printout/0,9788,109642,00.html> (accessed 5 December 2001).
39 On Taiwan's paradox, see M. Landler, 'Taiwan's PC Makers Shift to China', 29 May 2001. Online. Available HTTP: <http://www.taiwansecurity.org/NYT/2001/NYT-052901.htm> (accessed 12 March 2001).
40 On Stan Shih's comment, see Reuters, 'Vice President Annette Lu Seeks to Cool China Fever', 18 January 2001. Online. Available HTTP: <http://www.taiwansecurity.org/Reu/2001/Reuters-01190t.htm> (accessed 12 March 2002).
41 This much-disputed consensus was reached in negotiations between the two sides held in Hong Kong in October–November 1992.
42 On the hi-tech zones and development in Shenzhen, see T. Saywell, 'Watch Your Back', *Far Eastern Economic Review*, 16 September 1999, pp. 52–4; and also Li, Ning 'High-Tech Economic Zones: New Impetus Pushing Economy Up', *Beijing Review*, 24 April 2000, p. 15.
43 M. Miller, 'CyberCity's High Ambition', *South China Morning Post*, 28 September 1999. Online. Available HTTP: <wysiwyg://77/http://www.scmp.com/News/...xt_asp_ArticleID-19990928014602493.asp> (accessed 4 October 1999).
44 On the relationship between CyberCity and Fraser and Neave Ltd, see F&N, 'Ascendas and CyberCity Join Hands in Consortium to Take Over Hing Kong', 5 March 2001. Online. Available HTTP: <http://coldfusion.ascendas.com/news_view.cfm?PID=14> (accessed 27 February 2002).
45 These tendencies for competition within the Greater China region are further fuelled by similar projects elsewhere in Asia, such as Malaysia's Multimedia Supercorridor and Singapore's 'Intelligent Island'.
46 L. Malvezin, 'Asian Affairs Interview with Li Zibin, Former Mayor of Shenzhen – Vice-Minister, Hong Kong Needs Shenzhen, Shenzhen Does Not Need Hong Kong', *Asian Affairs*, Spring 2000 Issue, p. 4. Online. Available HTTP: <http://www.asian-affairs.com/China/lizibin.html> (accessed on 21 March 2001).
47 L. Malvezin, 'Asian Affairs Interview', p. 4.
48 On the strategic dilemma, see M. Landler, 'Taiwan's PC Makers', p. 2.

6 What's in a name?
China and the Domain Name System

Monika Ermert and Christopher R. Hughes

By taking different political perspectives on China's 'digital leap forward', the chapters in this book assume that there is no such thing as a socially neutral technology. However, while it is evidently true that the social impact of a machine like the Internet depends on the ways in which it is appropriated by particular societies, it is also important to avoid the extreme position of assuming that artifacts can be molded to fit political purposes without any limitations imposed by their technical specifications. The case of Robert Moses the New York builder might demonstrate that a bridge can be used to divide people just as well as it can be made to connect them,[1] yet it is also possible to find examples of technologies that seem to be 'inherently political' in that they demand the formation of certain kinds of political systems if they are going to be used effectively. The classic example is nuclear power, the safe use of which demands a significant sacrifice of civil liberties, through measures such as increasing the use of background security checks and covert surveillance in order to prevent certain materials falling into the hands of terrorists and other criminals.[2]

In the case of ICTs, the Domain Name System (DNS) might be just such an inherently political technology. It is certainly a significant source of political power due to its function of allocating, storing and retrieving Internet addresses. Yet its inherent political characteristics also stem from the degree of centralization that has to be built into the technology if it is to ensure technical standardization and the maximum potential for interconnectivity between systems and avoid making multiple allocations of the same addresses. Whoever exercises control over this centralized technology, however, inherits the power to decide who exists in cyberspace and under what identity.

Due to the historical development of the Internet, both the central technology and the management system of the DNS have come to reside in the United States. Countries such as the PRC, which joined the system relatively late, are thus faced with the problem of trying to establish some kind of control over a system that directly affects what they regard to be their rightful portion of cyberspace. By looking at how this situation has come about, this chapter will present the DNS as a case study of how a certain kind of technological development seems to determine certain kinds of decision-making structures, which have in turn led to international tensions between the PRC and the United States.

Politicization of a technical system

The technical job of the DNS is to manage the way in which Internet addresses are organized, stored and retrieved. To be efficient, this requires the ability both to allocate addresses according to universally accepted standards and to ensure that such addresses are not duplicated by the many machines that are connected to the Internet for the provision of content and services. Its origins can be traced back to the early 1980s, when Internet administration was still the preserve of a small number of professional computer engineers and standardization seemed to be no more than a technical and organizational issue to be resolved as the number of computers connected to the Internet rose. Today, this system forms the foundation of the global address mechanism upon which the functioning of the Internet depends.

The basic principles of the DNS were put in place by engineers based in the United States, such as David Mills, Jonathan Postel, Zaw-ming Su and Paul Mockapetris in the early 1980s. They developed the idea of using mnemonic tools as a substitute for unwieldy Internet protocol (IP) addresses that consist of long strings of digits. An IP address like '123.45.67.891' could thus appear as something more meaningful such as 'www.myuniversity.ac.uk'.[3] By 1985 this principle had led to the formation of the DNS standard, which had become widely adopted by 1987.

The need to guarantee interoperability, however, also determined that the DNS should evolve into a centralized system based on an 'A-Root Server' in which all the TLDs of the official 'root zone' have to be listed. This information is fed to twelve 'slave' root servers on a daily basis. When a request for an address is made, these root servers can then direct the inquirer to the authorized administrator for the relevant TLD. Some of these are private firms like the United States-based VeriSign Inc., which administers '.com' addresses. Others have a closer relationship with government, such as China's CNNIC, which administers '.cn'. Queries then travel to a local DNS server until the information requested is obtained. Such a centralized system thus gives the operators of the A-Root Server considerable power regarding the ability to grant 'existence' and identity in cyberspace, a fact underlined by the shadowy lives of those alternative domain providers who try to circumvent the system.[4]

The DNS is a remarkable achievement when one considers the special kind of 'legislative process' that produced it. The ideas of engineers like Postel and Mockapetris were developed in the form of documents known as 'RFCs', meaning 'Requests for Comment'. Solving technical problems through the circulation of RFCs began as early as 1969, when engineers were struggling to create common standards to exchange information between what at that time was little more than a handful of computers distributed throughout the United States. Such figures formed themselves into what became known as the Internet Engineering Task Force (IETF), which describes itself as 'unusual in that it exists as a collection of happenings, but is not a corporation and has no board of directors, no members, and no dues'.[5] This informal style of organization and negotiation grew out of the fundamentally libertarian ethos of the engineers involved. This was later to underpin what is known as the 'Open Source' movement, based on the principles that nothing should be kept secret, problems should be solved through collaboration, and all results should be in the public domain.[6] In practice, the RFC process proved to be remarkably efficient, as memos were sent out for comment when a technical problem arose, and recommendations were adopted as they gained broad support. Proposals that passed the IETF process were widely respected by the community of developers, who realized the necessity to agree on common standards that could ensure the interoperability of the system.

Great strains were placed on this style of governance by the spectacular growth of the Internet that occurred under the impact of commercialization in the 1990s, however. The creation of the World Wide Web by Tim Berners Lee in 1991 and the bestowing of authority on the National Science Foundation of the United States by Congress to allow commercial activity on the Internet the following year opened the way for the Internet to become the mass means of communication that we know today. As the number of applications for domain names rose dramatically, the job of allocation that had been handled by Jonathan Postel was contracted out to the private sector company, Network Solutions Inc. (NSI). The allocation of domain names became a profitable activity as NSI was allowed to start charging for its services, and the first 'dot.com' boom saw exorbitant prices being asked for short, easy to recognize addresses like 'www.business.com'.

The United States government further developed its response to the increasingly complex and commercialized nature of the DNS by contracting out its overall management to an organization called the 'Internet Assigned Numbers Authority' (IANA), which was really just the figure of Jonathan Postel himself. Meanwhile, technicians tried to keep up with the rapid growth of demand by frequently updating the RFCs.[7] Several 'Top Level Domains' (TLDs) were introduced, under which individual users could register their addresses. Some of these indicated specific functional constituencies, like '.mil' (for the United States military), '.gov' (for the United States administration), '.edu' (for universities, mainly in the United States), or '.int' (for international organizations). 'Generic Top Level Domains' (gTLDs) were also introduced for more general use, such as the well-known '.com', '.net', '.org', and the more recently introduced '.biz' or '.info'.

Alongside these gTLDs, IANA also created a system of 'country code Top Level Domains' (ccTLD). The names they used were derived from what is known as the 'Alpha-2 code elements' used by ISO standard 3166–1. This is the internationally accepted list of all the countries recognized by the UN in abbreviated form (such as '.cn' for 'China'). Going down this route allowed technicians like Postel to avoid interference with politics and getting involved in the sensitive issue of deciding what is and what is not a country. Giving a ccTLD to an entity like Taiwan, for example, could present technicians with a real political problem. However, the use of '.tw' has never been opposed by the PRC in the same way that it objects to the island joining international organizations requiring statehood, like the UN. The reason for this could be that according to ISO 3166–1, 'tw' stands for 'Taiwan, province of China'.[8] But it allows Taiwan to have an 'independent' identity on the Internet and the two Chinas to live side by side in cyberspace, with their respective managers even having friendly relations.

As the Internet grew in size and complexity under the impact of e-commerce, however, it became increasingly important to ensure that management of the DNS was kept at arms length from interference by the United States government. The solution sought by the Clinton administration was to privatize the service by creating the non-profit-making Internet Corporation for Assigned Names and Numbers (ICANN), which was established under California law in October 1998. From Washington's perspective, a private corporation seemed to be the best way to maintain a centralized system of administration on United States soil that was not under its own control.[9] Yet ICANN has been seen by many as amounting to little more than an oversight body working for the United States Department of Commerce, ensuring the continuation of the American dominance over DNS governance that can be traced back to the early engagement of United States public institutions in the organization of the Internet.

The technical nature of the DNS makes it hard to see how such a situation can be avoided. The most obvious problem is that ICANN is ultimately subservient to the United States government because the A-Root File is at the heart of the DNS, and any changes to it have to be approved by the Department of Commerce. ICANN thus has to seek approval from the government for any modification, such as creation and changing country zones like '.cn', with all such measures requiring the countersignature of an official of the United States National Telecommunications and Information Association (NTIA), a subdivision of the Department of Commerce.

Concerns have also been raised over the fact that ten out of the thirteen DNS root servers are operated by United States' institutions and companies, the remaining three being in Stockholm, Tokyo and Amsterdam. The fact that United States firms also dominate the market for the provision of domain names is also largely due to the fact that the DNS was developed on American soil. The biggest player is thus VeriSign, which acquired NSI in 2000 and which is also contracted by ICANN to manage the A-Root Server on its behalf, under supervision by the Department of Commerce. Finally, the fact that ICANN is established as a Californian corporation raises the issue of whether the registrar companies which it accredits to administer domain names around the world, come under United States jurisdiction.

Such developments have led many critics of ICANN to conclude that there needs to be more international participation in the decision-making process, especially concerning initiatives such as the selection of new gTLDs, like '.info' and '.biz'. Yet moves to 'democratize' ICANN so far have been rather farcical. As a private corporation, it is governed by nineteen directors. An attempt to make this board more democratic was made when nine of these positions were made into 'at large' representatives of five 'world regions', and were elected by on-line ballot in October 2000. The winner of the 'at large' directorship to represent all Internet users in the Middle East, Pakistan, India, China, Japan, Australia, Afghanistan and 'countries to the East', including the East Indian Ocean islands and Antarctica (but excluding United States and Latin American possessions) was the Maryland-based Japanese employee of Fujitsu, Masanobu Katoh, who polled no less than 13,913 votes!

An attempt to satisfy the demands of states to have an input into the decision-making process was made by creating a General Advisory Committee (GAC), which consists of representatives from thirty-two concerned national governments, including the PRC. This body has issued principles which attempt to reassert the authority of states, by requiring that any domain name registrar should only be approved by ICANN after a communication has been received from the host government to authorize its existence, and that if an administrator does not have the support of the local community and authorities then its licence should be reallocated to another delegated body. Ultimately, however, the concentration of power that is decided by the technical nature of the DNS means that there are can be no binding force to make such changes to the management system of ICANN effective in achieving a more representative system of governance.

The answer is thus sought by some critics of ICANN in the development of a more decentralized technical system. In Europe, for example, discussions are going on about the development of a parallel root server system.[10] In Asia, one of the most outspoken advocates of a decentralized method for overseeing DNS administration is Tan Tin Wee, Associate Professor of Biochemistry at the University of Singapore.[11] Tan is chairman of the Multilingual Internet Names Consortium (MINC) which has gone beyond Asia to find partners in the Arab and African worlds. Several of his students were also prime movers in the establishment of iDNS.net, a spin-off enterprise of the University of Singapore that carries out research on Chinese domain names and has become a key pressure group in the growing movement for the internationalization of the DNS.

China enters the DNS

China itself entered the DNS at a comparatively late stage. The first network link between China and the outside world was established on 20 September 1987 when the Chinese Academic Network (CANET) was connected to the Internet through cooperation between the Institute of Computer Application (ICA) in Beijing and the University of Karlsruhe.[12] It was not until 1990, however, that the ccTLD '.cn' was delegated by Jonathan Postel to Professor

Qian Tianbai, then deputy chief engineer at ICA and manager and administrator of the CANET. In 1993 CASNet, operated by the Chinese Academy of Sciences, began to study the DNS, and the following year Qian Tianbai started to actively manage the '.cn' space for China, again with assistance from the University of Karlsruhe. It was not until 1994, though, that China's first Website, 'Window on China', went online.

According to CNNIC's account of this process, one of the reasons for China's late entry into the system was that the National Science Foundation of the United States rebutted requests for an 'official' connection to the Internet several times during 1992 and 1993. This, claims the CNNIC, was because Washington opposed any socialist countries gaining access to the Internet, due to the amount of scientific and technological information it contained and the number of its own official institutions that had already gone online.[13] From the American perspective, however, the delay was due largely to the fact that the Chinese had not yet gained a proper understanding of how the Internet works. Cindy Zheng of the San Diego Super Computer explained after a visit to China in 1993, for example, that while the United States did impose restrictions on the export of high-end computer systems to China, additional difficulties were created for joining the DNS by other factors. For example, the Chinese did not realize that they should approach commercial carriers and network providers for a positive response to their requests, rather than government officials. The way in which networking projects in China were jointly funded by the World Bank and the Chinese government also caused problems. Of special importance, though, was the high degree of distrust that existed between the different institutions taking part in the National Computer Networking Facility of China, which brought together the main academic research institutions involved in computer sciences.[14]

Such distrust between Chinese institutions may have been partly due to the general situation that existed before computer networking began to be regulated in the mid-1990s. As in other countries, academia had been the original driving force behind the Internet in China, but funding had been sparse and partly derived from foreign research institutions. When the need for coordination grew with the rising number of institutions setting up their own networks and links to the outside world, there were no existing models for cooperation, and ICANN's practice of consensus-based self-regulation was viewed with suspicion. As the 1994 INET Report on Networking in China makes clear,[15] the result was a situation in which too many ministries and other authorities were involved in decision (or non-decision) making. These included the State Science and Technology Commission, the Ministry of Posts and Telecommunications, the Ministry of Education, the Ministry of Electronic Industries, the Beijing Posts and Telecommunications Administration, and other regional and provincial authorities. According to the INET Report, there was no agreement among these bodies about the importance of academic networking and how to pay for the development and operations of the various networks.

Such a situation may well have been a primary reason for the slow progress of the setting up of the '.cn' DNS registry. In this situation of bureaucratic competition, no agreement could be reached on who should be in charge of the assignment and management of domain names, and how the domain name server should be set up and managed. As Cindy Zheng points out, different opinions arose over issues such as whether the naming system under '.cn' should be organized according to a geographical substructure, or with second-level generic domains according to the existing gTLD scheme, such as '.com.cn', '.org.cn'. The settlement of such disputes had to wait until a regulatory system had been established and a degree of centralized authority had been created in the form of the CNNIC, a non-profit organization wholly owned by the state, and controlled by the Informatization Group of the State Council.

The establishment of the CNNIC, however, meant that domain name registration had effectively become a function of the Chinese state. Although there are no clear rules by which the organization maintains the 'purity' of the various second-level domains, it claims the right to administer not only the Chinese namespace under the ccTLD '.cn', but also to require notification from operators of servers in China using any other domain. Domain registrants have to be institutions or companies, and not individuals, while foreigners have to have residential status. All have to be able to produce a document of verification from the organization for which they work. The CNNIC is also unique in the world for the way in which it carries out a *manual* check of applicants' documents. The workload as a result of that procedure may provide some explanation for the comparably high annual registration fee of RMB 300 per domain that is set by the CNNIC.

Although it is hard to assess just how many of the CNNIC provisions are effectively executed, the Chinese authorities have at least established the principle that it is they who have the right to maintain a tight grip on what they consider to be their rightful namespace. The state has also expanded its control over the CNNIC structure through a number of measures that reduce the input from academics. When the CNNIC was established, for example, the research institutions that had previously been engaged in DNS administration were absorbed into a kind of oversight body, known as the 'Steering Group', which meant that the Chinese Academy of Sciences (CAS) lost its direct influence over day-to-day management. This movement away from academic involvement continued when the MII was established and took over the job of supervising the CNNIC from the Informatization Group of the State Council. Under the new MII regime, moreover, the CNNIC Steering Committee was reorganized to bring in commercial telecom players, like China Telecom, as associate members.

Despite the bureaucratic politics surrounding management of the DNS in China, however, the number of domains registered has grown dramatically as uptake of the Internet has increased. While there were only two hosts in the '.cn' domain two years after its management was delegated to Qian Tianbai in 1990, and around 1,000 before the establishment of the CNNIC, by the end of 1997 the number had risen to 5,000. Today the figure has reached 127,319. The majority of those that come under the '.cn' domain are registered as '.com.cn', which means that they are commercial organizations (see Figure 2.4, p. 36). However, Chinese users can also register under the gTLD '.com', which makes the ccTLD '.cn' redundant. It is impossible to put a figure on how many users do this, but there are certainly registrars who sell '.com' domains in China at a cheaper price than '.com.cn'. It is worth noting, moreover, that there is no official figure for the number of foreign companies that have registered under '.cn' either.

Clash of systems

With both the United States and the Chinese governments attempting to exert control over an increasingly commercialized DNS, it was inevitable that frictions would arise between them. This was especially so when ICANN decisions began to have a direct impact on the nascent market for selling domain names in China when the first 'dot.com' boom took off in major cities. Indigenous telecommunications providers, such as the Zhejiang-based Eastern Communications Co. (Eastcom), had in fact been quick to make efforts to enter the international market as operators of new gTLDs, well before other Chinese firms began to prepare for WTO accession. As part of the strategy to gain a share of the market, Eastcom successfully applied to become an ICANN accredited domain name registrar.

Eastcom quickly came under pressure from the United States, however, when a Virginia court ruled that the Hong Kong and Shanghai-based company Maya should relinquish the domain name 'CNNews.com' to AOL-Time Warner subsidiary CNN. Maya, however, had registered its claim to the CNN domain name with Eastcom. ICANN's legal counsel and vice-president, Louis Touton, ordered Eastcom to comply with the ruling by the United States court. According to the court protocols, the judge also considered ordering VeriSign/NSI, the central '.com' registry, to cancel the domain registration lodged by Eastcom. Although he explained that the court had no power to order anybody in China to do anything, he also reasoned that it had jurisdiction over NSI.[16] The implication of this threat to impose a cancellation order on VeriSign/NSI, or on a non-United States based registrar like Eastcom, is that anybody registering a domain name comes under United States jurisdiction, regardless of whether they go through a Chinese, German or South African ICANN accredited registrar.[17] Not surprisingly, the dispute over 'CNNews.com' led Maya to warn the Americans,'Yankee, don't bully people too much'.[18] Such sentiments resonated well with general criticisms in the Chinese press of what are seen as the ambitions of the United States to exert its hegemony in cyberspace.

Further frictions between the United States and China have also been created over the introduction of standards for the use of Chinese-language domain names. This problem arises because the DNS was originally limited to using the American Standard Code for Information Interchange (ASCII), which consists of the Roman alphabet and a number of graphic symbols. While most western countries are satisfied with this choice, even though they are denied the use of national language characters like the German umlaut and French accents, the call for nationalized versions of domain names has grown in many Asian and Arabic countries. The first trials of Chinese domain names were started in the late 1990s by Singapore's iDNS.net, as Tan Tin Wee argued that a reform of the DNS had to follow on the sinicization of software and e-mail and pushed for rapid movement towards the adoption of an internationalized system.[19]

In China itself, the CNNIC began to conduct research on developing domain names using Chinese characters in 1998 and in January 2000 it started to carry out tests using software that could impose a Chinese character address on top of an English address. This project expanded to take in the 'four territories and two coasts' (Mainland China, Hong Kong, Macao, Taiwan) when a Chinese Domain Name Consortium was established on 19 May 2000. Complaining that using English addresses on the Internet is like using an English-language map in a Chinese city, participants in this venture were particularly concerned that being forced to use English addresses for e-commerce would impose real disadvantages on established Chinese brand names.[20] Yet such a movement provoked concerns at ICANN, despite assurances that the new consortium would meet international standards by working with international organizations.

It is in this context of growing international frictions that the American firm VeriSign was accused of 'infringing on China's sovereignty'[21] when it started to create its own technical standards for Chinese domain names. In the autumn of 2000 VeriSign announced that it would start registering non-ASCII domains, including Chinese, first on a trial basis and using a technical standard that they thought would be the most viable within the ongoing discussion in the IETF.[22] The CNNIC reacted quickly by demanding a stop to VeriSign's activities. Its director, Ma Wei, explained to a press conference in Hong Kong that 'We hope that Chinese people would have the mandate over Chinese domain names'.[23] However, its main argument against VeriSign's initiative was more sophisticated, and relied on complaining about the fact that no international agreement had yet been reached on a common technical standard for Chinese names, so the IETF and ICANN should put pressure on operators to wait until global compatibility could be ensured. The Chinese representative in ICANN's GAC thus urged the board of directors to stop VeriSign from leaving its testbed environment and moving towards full-scale registration of Chinese domains. The GAC responded to such pressures by issuing a communiqué listing nine principles to constrain operators from leaving their testbeds without first gaining the general consent and coordination of a community-based framework such as ICANN.[24]

When ICANN proved reluctant to stop VeriSign grabbing another part of what CNNIC and Chinese operators regarded to be their rightful market, the result was further alienation from the American-centred Internet governance system and even greater scepticism towards the process of bottom-up standardization. Chinese officials have criticized initiatives such as those taken by Verisign for being a kind of linguistic hegemony. As Minister for Information Industry Wu Jichuan put it to the Pacific Telecommunications Conference:

> Due to historical and technical reasons, 90 per cent of the information available on the Internet is in English and the overwhelming majority of it is generated from developed countries, whereas developing countries are mostly information receivers. As information flows across borders and developing countries are absorbing advanced technological and cultural information, their cultural traditions, moral standards and values have been severely challenged.[25]

An organization like CNNIC thus sees sinicization of the DNS to be one of its main goals in a battle to prevent American technological and cultural dominance of the Internet.

Conclusion

Bearing in mind the common perception that the Internet is a technology that erodes the power of the state, it is somewhat ironic that one consequence of the informal development of a technically centralized DNS structure has been the concentration of administrative power in the United States. This chapter has shown how this situation has created political tensions between states, as governments around the world have tried to exert their influence over what has become a central institution of Internet governance. China has been a party to such disputes partly because it joined the DNS relatively late and then felt it necessary to take measures to regain control over the address system which had a direct impact on its commercial activities. Domestically, this meant the state winning control from the academics and engineers who were originally entrusted to run the system. Internationally, it required responses to measures taken by the United States government that appeared to consolidate the centralization of decision-making power under American jurisdiction.

A further twist in the story appears with the corresponding growth of opinion in bodies like ICANN's GAC that domain name registration should now be recognized as an official function of states. Within China, the government has taken its own measures to challenge the technological status quo by putting in place bureaucratic and regulatory structures to exert control over its ccTLD. While China is lobbying for acceptance of this model by the international community, however, it can do little about the fact that the technical structure of the DNS concentrates real power in the A-Root Server, which is located on American soil and is ultimately under the control of the

United States Department of Commerce. Moreover, the technological capabilities of American corporations like VeriSign, means that even indigenous Chinese attempts to define standards for a Chinese-language name system are being challenged from across the Pacific. The DNS, as it has developed so far, therefore, appears to be an inherently political technology that determines the formation of a centralized power structure. Any movement away from this will depend as much on the development of new standards and architecture as it will on the ability of states to put in place a more satisfactory system of global governance.

Notes

1 R.A. Caro, *The Power Broker: Robert Moses and the Fall of New York*, New York: Random House, 1974. See also L. Winner, 'Do Artifacts Have Politics?' in D. Mackenzie and J. Wajcman, *The Social Shaping of Technology* (Second Edition), Buckingham and Philadelphia: Open University Press, 1999, p. 30.
2 L. Winner, 'Do Artifacts Have Politics?', pp. 33–8.
3 T. Denton, 'The Governance of the Domain Name System'. Online. Available HTTP:<http://www.tmdenton.com/Speeches/The%20Governance%20of%20the%20Domain%20Name%20System.htm> (accessed 5 April 2002).
4 There are several alternative TLD operators, such as Image Online Design or the various organizations that are grouped under the Open Root Server Confederation (ORSC). None of these, however, is able to attract large numbers of ordinary Internet users.
5 IETF, 'The Tao of the IETF: A Novice's Guide (RFC 3160)'. Online. Available HTTP: http://www.ietf.org/tao.html#intro (accessed 21 May 2002).
6 On the RFC process see John Naughton, *A Brief History of the Future: The Origins of the Internet*, London: Weidenfeld and Nicolson, 1999, pp. 137–9.
7 Most notable are RFC 1034 and RFC 1035, written by Mockapetris, which superseded earlier drafts. Details of DNS-related RFCs can be found on the IANA Website. Online. Available HTTP: <http://www.iana.org> (accessed 1 June 2002).
8 For the full list of the Alpha-2 code elements of ISO 3166-1 see 'English Country Names and Code Elements'. Online. Available HTTP: <http://www.din.de/gremien/nas/nabd/iso3166ma/codlstp1/en_listp1.html> (accessed 21 May 2002).
9 United States Department of Commerce, 'Management of Internet Names and Addresses'. Online. Available HTTP: <http://www.ntia.doc.gov/ntiahome/domainname/6_5_98dns.htm> (accessed 1 June 2002).
10 Andy Mueller-Maguhn, 'elected' as the so-called ICANN 'at-large Director' for Europe has put an independent European root zone on the ICANN agenda several times.
11 See proceedings of the ITU conference on Multilingual Domains, especially Tan Tin Wee, 'Policy and Coordination Issues in Multilingual Internet Names'. Online. Available HTTP: http://www.itu.int/mdns/presentations/dayone/tan1.ppt (accessed 30 May 2002).
12 CNNIC, 'Evolution of the Internet in China'. Online. Available HTTP: <http://www.cnnic.net.cn/evolution.shtml> (accessed 9 April 2002); B. McIntyre, 'China's Use of the Internet', in P. Lee (ed.), *Telecommunications and Development in China*, New Jersey: Cresskill, 1997, pp. 159–69.
13 CNNIC, 'Evolution of the Internet in China'.
14 Cindy Zheng, 'A Special Report – Current Computing/Networking Status in China', China News Digest, Special Issue on Networking in China, 11 July

1993. Online. Available HTTP: <http://www.sdsc.edu/~zhengc/93trip.html> (accessed 12 May 2002).
15 F. Kuo, J. Ding, C. Zheng, F. Hussain, 'Issues in Academic Networking in the PRC: INET 1994 Report', San Diego Supercomputer Centre Website. Online. Available HTTP: http://www.sdsc.edu/~zhengc/inet94.html (accessed 30 June 2002).
16 Court protocol available online. Available HTTP: <http://www.cptech.org/ecom/jurisdiction/CNNEWS.pdf> (accessed 1 June 2002); <http://www.cnnews.com/topic/388.shtml> (accessed 1 June 2002).
17 In addition to '.com', the registries for many other gTLDs are also located in the United States. For example, VeriSign also has the official registry for '.net', '.org' and '.edu'. The registries for the two newest gTLDs, '.biz' and '.info', both have offices in the US in addition to Australia and Ireland. Thus, under the Virginia court's interpretation of the Anticybersquatting Consumer Protection Act (ACPA), every domain name that is registered under '.com' or any of the other above-mentioned gTLDs, is subject to US jurisdiction. See V. Polak, W. Matus and S. Gelin, '".com" Domain Names can Lead to U.S. Jurisdiction', *Internet Law Journal*, 2 February 2002. Online. Available HTTP: <http://www.tilj.com/content/litigationarticle01300201.htm> (accessed 10 April 2002).
18 Cnnews.com, 'Meiguo lao, bu yao qi ren tai shen', 12 April 2001. Online. Available HTTP: <http://www.cnnews.com/maya/cnnews/zt/ztwz/item/2001_04/486780.shtml> (accessed 1 June 2002).
19 International Telecommunications Union, 'Joint ITU/WPO Symposium: Creating a Wider Understanding of the Complex Issues Surrounding the Implementation of Multilingual Domain Names', 22 February 2002. Online. Available HTTP: <http://www.itu.int/itunews/issue/2002/01/joint.html> (accessed 31 May 2002).
20 CNNIC, '*CNNIC tuichu zhongwen yuming shiyan xitong*' ('CNNIC Promotes Experimental Chinese-Language DNS'). Online. Available HTTP: <http://www.cnnic.net.cn/cdns/about_cdns.shtml> (accessed 1 June 2002).
21 Reuters, 'China Claims Its Own Domain'. Online. Available HTTP: <http://www.wired.com/news/politics/0,1283,40506,00.html> (accessed 1 June 2002).
22 For details on the testbed see VeriSign, 'General Information Paper on Internationalized Domain Name Resolution', 3 April 2001. Online. Available HTTP: < http://www.verisign-grs.com/idn/Gen_Info_Paper.pdf> (accessed 31 May 2002).
23 Reuters, 'China Claims Its Own Domain'.
24 ICANN, 'Communique of the Governmental Advisory Committee, March 10, 2001', and 'Communique of the General Advisory Committee, June, 3rd 2001'. Online. Available HTTP: <http://www.icann.org/committees/gac/communique-10mar01.htm> (accessed 1 June 2002); <http://www.icann.org/committees/gac/communique-03jun01.htm#Attachment> (accessed 1 June 2002).
25 Roman Rollnick, 'China Concerned at Electronic Threats to Moral Standards.' Earth Times News Service, 14 January 2002. Online. Available HTTP: http://www.earthtimes.org/jan/telecommunicationchinajan14_02.htm. For a discussion of security issues stemming from dependence on international networks see Chapter 7 by Christopher R. Hughes in this volume.

7 Fighting the smokeless war
ICTs and international security

Christopher R. Hughes

Several chapters in this volume have shown how the Chinese leadership places ICTs at the heart of the state's economic development strategy. Despite attempts to create an indigenous information industry sector, however, the hard reality remains that the core research and development upon which technologies like the Internet depend, as well as key administrative institutions like ICANN,[1] are based in the United States.[2] The emergence of such a degree of dependence is stimulating intense debates over the relationship between ICTs and international security in the PRC, in which the discussion revolves around a set of problems that can loosely be grouped under the heading of 'information warfare'.

Some writers trace the origins of information warfare back to the ancient Chinese strategist Sun Zi, who advocated achieving victory through deception, knowing the mind of the enemy and gathering intelligence.[3] History is in fact littered with examples of information determining the course of war. The United States may never have entered the First World War if the British had not intercepted and deciphered the 1917 Zimmerman Telegram that revealed a plan for the Germans to ally with Mexico.[4] The increasing sophistication and efficiency of ICTs, however, combined with the global integration of networks, creates new potentials for information warfare which dramatically magnify its importance as a strategic threat.

In one respect this can be seen in the new possibilities for projecting what Keohane and Nye call 'soft power', the ability to establish norms, institutions and agendas through propagating one's own culture and shaping the preferences of others.[5] Yet the spread of the Internet has also made possible new kinds of aggressive measures that can seriously disrupt important social infrastructures. These include actions such as triggering data overload, spamming and attacking software with viruses, Trojan horses and 'worms'. It is also possible for manufacturers and engineers to leave hidden trapdoors in systems that make unwarranted surveillance possible. Great damage and inconvenience can also be caused by 'hacking', and even by physical attacks by electromagnetic pulse, electronic countermeasures, and conventional military strikes.[6]

The implications of such developments for international security are still open to wide-ranging debate in military circles throughout the world. On the one hand, radical proponents of what has come to be known as the Revolution in Military Affairs (RMA) propose that future conflicts will be decided by 'cyber warfare', aimed at disrupting, disabling or exploiting critical information nodes, and 'net war' that involves deception and psychological operations to influence the behaviour of the enemy through deterrence and the shaping of perceptions.[7] On the other hand, sceptics claim that information warfare presents more of an Achilles' heel for the technologically advanced societies in which both military and civilian critical infrastructures are highly dependent on digital networks, raising the spectre of an 'electronic Pearl Harbour',[8] or an 'electronic Waterloo'.[9] Some commentators even suggest that technological backwardness and authoritarian politics might in themselves offer protection against information warfare. As Henry and Peartree put it, 'What use will niche-casting propaganda be against an enemy leader who does not have satellite television or an Internet connection? In 1997, half of the world's population had never even made a telephone call'.[10]

The PRC, an authoritarian developmental state going all out for siliconisation, thus presents an interesting case study. While the country is underdeveloped by many standards, the rate of connectivity among the elite urban population is far from politically insignificant. President Jiang Zemin himself has a personal Internet connection and logs on regularly. Such a situation thus presents Chinese policy-makers with something of a double-edged sword when it comes to considerations of security. While the technological lag behind potential adversaries, such as Japan, Taiwan and the United States, makes the country vulnerable to attack, new opportunities are also arising for launching information warfare against societies that have a very high degree of dependence on digital networks. This chapter will assess how policy-makers in China are responding to such a dilemma by looking at the debate across a broad range of organisations, including the military, state ministries, the CCP and academia.

Fighting the smokeless war

From an American perspective, the ability for new technologies to carry ideas across China's borders is not necessarily a bad thing. Secretary of State James Baker began to develop this theme at the end of the Cold War, when he explained, 'It is in our interests that the next generation in China be engaged by the Information Age, not isolated from global trends shaping the future'.[11] Vice-President Al Gore was even more upbeat when he launched the Global Information Infrastructure (GII) project in March 1994, explaining that, 'To promote . . . to protect . . . to preserve freedom and democracy, we must make telecommunications development an integral part of every nation's development. Each link we create strengthens the bonds of liberty and democracy around the world.'[12] Secretary of State Madeleine Albright also claimed that the accelerated development of the Internet and telecommunications in China after its accession to the WTO would have an impact on the human rights and political situation by increasing contact with Americans and other foreign trading partners,[13] and reducing the power and reach of government censorship.[14]

From the point of view of the CCP leadership, though, the propagation of American culture and values inside China is part of what Deng Xiaoping called the 'smokeless war' to undermine the socialist system through a process of 'peaceful evolution'.[15] Chinese academics thus warn about the ways in which ICTs can destabilise politics by making it easier for new actors to organise themselves and challenge the status quo, strengthen Western pressures for 'global governance', and lead to the development of transnational organisations and structures that challenge the maintenance of 'information borders' and 'information sovereignty'. Concerns are also expressed over the ways in which the new international regimes that are emerging to regulate the use of ICTs tend to be determined by the degree of hard and soft power that the most technologically advanced states are able to exert. Meeting the consequent challenges to state sovereignty is made even more complex, they claim, because the Internet magnifies the sources of post-Cold War instability in areas such as financial markets, environmental problems, terrorism, and non-military intervention in the domestic affairs of other states.[16]

Such concerns are located within a comprehensive analysis of the impact of ICTs on national security contained in a joint report by the Ministry of Information Industry (MII) and the CCP's Central Policy Research Office. This proposes that the threat of information warfare should be understood within a broad vision of global power that is based on an up-dated version of Mao Zedong's theory of the 'Three Worlds'. Just as Mao believed that the world was divided into three tiers of states, with the superpowers at the top, the developed states in the middle and the developing states at the bottom, in the information age there is also supposed to be three types of state. At the top of the pile is the 'information hegemony state', asserting its control by dominating the telecommunications infrastructure, software development, and by reaping profits from the use of information and the Internet. After this comes the 'information sovereign state', exemplified by those European states that have accumulated sufficient know-how to exert independent control over their information resources and derive profits from them, and to protect themselves from information hegemony. At the bottom of the pile are the 'information colonial and semi-colonial states', which have no choice but to accept the information that is forced on them by other states. They are thus left vulnerable to exploitation because they lack the means to protect themselves from hegemonic power.[17]

According to this theory, the present international situation has already revealed how certain states can combine their traditional military and economic advantages with their lead in information technology in order to contain the development of the PRC, exploit its resources, destroy its culture, and attack its politics, military and economy. By waging psychological warfare through e-mail and electronic newspapers, wreaking destruction by leaving Trojan horses and viruses in software that is sold to China, and by leaving 'back doors' in hardware, the technologically advanced states can obtain advantages that they cannot gain through military means.[18] The overall result is a kind of virtual realism, in which the survival or death of the PRC and its ability to take initiatives in the struggle for development depends on whether it can consolidate and expand its 'information territory' and preserve 'information borders', defined not by geography but by the scope of politically influential information and the building of strong 'spiritual defences'.[19]

The armed forces are equally concerned about the threats posed to national security by information warfare. Among their fears are developments such as the provision of serial numbers for Intel processors since the launch of the Pentium III, which could allow foreign powers to identify users and provide them with access to all kinds of possibly sensitive information. Similarly, Microsoft operating systems since the launch of Windows 98 are viewed with suspicion due to their ability to interact with hardware and generate a code related to the user's name and address which can be transmitted to the Microsoft Website. Viruses are also a cause for concern. The CJH virus, for example, is claimed to have caused Chinese enterprises an overnight loss of more than RMB1 bn at one point in 1999. The military also believe that the United States Secret Service disabled Iraqi air defences during the Gulf War by installing chips with a virus in computer systems that Iraq had acquired from France, which could be triggered by remote control.[20]

Debates in the main military newspapers thus show a high degree of concern over the implications of ICTs for military strategy and doctrine.[21] The vulnerablity of support systems to information warfare, for example, is a prominent theme in articles on military logistics published in 2001 and brought together in a special edition of the online 'Military Affairs Salon' of the *People's Liberation Army News*.[22] The consensus on this topic is that informatisation is essential in an age of increasingly mobile, multi-theatre, integrated warfare. As one report in the *People's Liberation Army News* points out, for instance, under the Ninth Five-Year Plan large amounts of capital were invested in informatisation of the military's medical service, with the establishment of over 50,000 Websites and some 38 model informatised hospitals.[23] Yet such developments are held to present the enemy with a growing number of soft targets. Meanwhile, Chinese commentators draw attention to how the Pentagon has continually upgraded the importance it attaches to its ability to conduct information warfare, evidenced by the emphasis given to information technology in its 2001 *Quadrennial Defense Review Report*,[24] and by the way in which it has been spending large amounts of capital on researching the use of viruses to disable and disrupt enemy computers since at least 1987.[25]

Building information borders

Faced by such threats, military analysts urge building what some have called an 'Internet Great Wall' (*wangluo changcheng*).²⁶ Part of this involves defensive measures like the development of decentralised, mobile and stealthy information systems rather than concentrated and large-scale IT structures. National security information in particular, it is argued, should be located in distributed and localised systems.²⁷ There is also an awareness that securing information borders against the possibility of information warfare requires the mobilisation of all the nation's military and civilian expertise. This is reflected, for example, in the linkage made by military leaders between the world Revolution in Military Affairs and Jiang Zemin's ideological campaign of the 'Three Represents', which advocates that the CCP should represent 'China's advanced productive forces, the orientation of China's advanced culture, and the fundamental interests of the overwhelming majority of the people in China'. General Fu Quanyou, Chief of Staff and member of the Central Military Commission, sees such an ideology as being compatible with the strategies and tactics of 'modern people's war'.²⁸

Such a view represents an interesting twist in the development of military doctrine that has been going on since the death of Mao Zedong in 1976. This originally moved away from the theory of 'people's war', according to which the enemy will be worn down by guerrilla tactics after penetrating deep into Chinese territory, towards one of 'active defence' along a 'strategic boundary' to stop an enemy before it can penetrate the country's external borders.²⁹ High technology took an increasingly central role in this doctrinal development thanks largely to international events such as the expulsion of Argentine forces from the Falkland Islands by the Royal Navy in 1982, but most spectacularly the 1991 Gulf War. It was this conflict that most dramatically revealed how the integration of ICTs with battlefield activity through satellite links, tracking and targeting systems, and airborne warning and control systems (AWACS), can enhance command, control, communications and intelligence (C3I) capabilities.

In 1993 the Central Military Commission thus announced a new doctrine of preparing to fight 'high-technology local wars under modern conditions' (*xiandai tiaojian xia de gaojishu jubu zhanzheng*). A further push towards high technology followed in 1997, just after the stand-off with the United States' Seventh Fleet during the Taiwan Strait crisis of the previous two years. It was then that the Central Military Commission announced the 'two basic changes' (*liangge jibenxing zhuanbian*) of: 'change from dealing with local wars under ordinary conditions to winning local wars under modern technology, especially high technology, conditions; change from an army of number and scale to an army of quality and efficiency, and from a manpower-intensive to a technology intensive army.'³⁰ President Jiang Zemin took up this theme when he made his report to the Fifteenth Congress of the CPC in September 1997 and promoted the professionalisation of the PLA in order to fight a defensive war under conditions of high technology and advocated the building of a 'strong technological army'.[31]

Within these preparations for high-technology warfare, the concepts of electronic and information warfare have been recognised as special types of campaign. According to testimony by an official from Taiwan's Ministry of National Defense, Beijing began to develop plans for information warfare as early as 1985, started to implement them in 1995, and began to conduct exercises using computer viruses to interrupt broadcasting systems and military communication systems in 1997.[32] By the late 1990s articles in the *People's Liberation Army News* were openly discussing tactics such as disrupting an enemy's communications systems through ensuring electromagnetic control, combining 'active interference with passive interference, electronic interference with repressive interference'.[33] They might have been encouraged in this thinking by incidents such as the 'Army After Next' winter war games held by the United States military during the late 1990s, in which more than 50 per cent of the home side's military information infrastructure was degraded by laser and electromagnetic pulse bomb attacks on its communications satellites as the mock battle began.[34]

Information warfare has certainly become a central theme in military manoeuvres. A national training campaign to create a 'strong technological army' was launched at the end of 1998, and information warfare was pushed still higher up the military agenda at the time of the 1999 NATO campaign against Serbia. Many of the military exercises held at this time involved online simulated combat, often between 'red' and 'blue' teams, with the scenario being a conflict over Taiwan or with the states neighbouring the South China Sea. Some, such as the exercises held in the Lanzhou military region in October 1998, focused specifically on electronic surveillance and counter-surveillance, disruption and counter-disruption, and destruction and counter-destruction measures. Special training corps for cyber warfare have also been established in some areas, such as the one established by an armoured division in the Nanjing Military Region to coach personnel in computer skills, software development and Internet warfare. Information warfare is also treated as a central element of combined-forces operations involving manoeuvres coordinated by advanced information systems, and tactics such as launching electromagnetic attacks to degrade the enemy's information systems. Teaching aids on information warfare are compiled by the various branches of the armed forces, drawing on experiences from wars fought by other armies around the world, using CD-ROMs to provide accounts of basic concepts, techniques and weaponry.[35] Researchers in Taiwan claim that the PLA has already reached a fairly advanced stage in its ability to use information technology to achieve command and control of the battlefield in any conflict with the island, and to launch an attack concentrating on soft targets such as computer networks used for banks, business and transportation.[36]

The broadening out of the 'technological army' to embrace expertise among the general population began to be developed most visibly by middle-ranking military cadres after the humiliation of the PLA by the intervention of the United States Navy in the Taiwan Strait crisis of 1995–6. The most notable example is the emergence of the doctrine of 'unlimited warfare', advocated by Qiao Liang and Wang Xianghui, both linked to the PLA Airforce Academy. Drawing broadly on a range of Western and ancient Chinese strategists, Qiao and Wang advocate defeating the overwhelming military power of the United States by using information warfare conducted via the Internet, combined with trade war and various permutations of terrorism, biological warfare, smuggling and the disruption of financial systems.[37] Such warfare would be 'popularised' (*pingmin hua*) in the sense that combatants would include teenage hackers as much as military professionals. Qiao and Wang are careful to point out, however, that the high degree of expertise required by such individuals distinguishes their doctrine from Mao's idea of 'arming the whole population'.[38]

Such views have fallen on fertile ground in the context of the upsurge of popular nationalism that was triggered largely by the Taiwan Strait crisis and further stimulated by the bombing of the PRC embassy in Belgrade in May 1999. There is ample evidence to show that the Internet was being used to launch information warfare from the PRC as early as 1998, when Indonesian Websites were targeted following the wave of atrocities committed against ethnic-Chinese Indonesians after the fall of the Suharto regime. Following the Belgrade incident, a much larger wave of activity took place against the Websites of NATO organisations, governments and political parties. Similarly, when the president of Taiwan, Lee Teng-hui, made a statement seen in the PRC as tantamount to a declaration of independence on 9 July 1999, over 7,200 attacks were launched against public Websites on the island.[39] Public Websites in Japan were also attacked in January 2000 when historians held a conference in Osaka questioning the historical truth of the Nanjing Massacre. At one point, some 1,600 strikes were launched within the space of seven minutes against the Bank of Japan's computer system.

This kind of cyber warfare seems to have been getting more organised. In part, this can be seen in a growing division of labour, according to which 'freshmen' attack mainly vulnerable commercial Websites under the guidance of more experienced hackers called 'knights'. There also appear to be organised groups of hackers in the making. A 'Chinese Hackers' Union' claimed to have gathered over 1,000 members within twelve days of the forced landing of a United States surveillance plane on Hainan Island after its collision with a Chinese fighter plane on 1 April 2001, who began placing Chinese flags and portraits of the missing Chinese pilot on United States Websites from 30 April onwards. Others quickly followed with similar actions, such as the 'Honkers Union' (literally 'red guests'), who claimed to have defaced some 700 United States Websites by the evening of 3 May. A portrait of Chairman Mao was the visiting card left by another group, composed of radical leftists calling themselves the 'Chinese Hawks' and known for earlier attacks on Websites such as those run by the religious Falungong movement.

Although such aggressive activity appears to be the result of largely spontaneous campaigns, there is some evidence to suggest that the state supports it at times. Successful hacking attacks against United States government computers after the Belgrade embassy incident, for example, were reported with a degree of pride in party-controlled newspapers, which printed the addresses of United States government Websites.[40] The Beijing municipal authorities even set up a special 'Sacred Sovereignty' Website on which people were encouraged to express their outrage over the Belgrade bombing, and from where they could obtain the e-mail addresses of NATO governments and political parties. Military commentators have also urged the establishment of 'information warfare brigades' that bring together expertise from across the whole spectrum of society, noting the example that has been set by the recruitment of hackers by states like the United States, India, the United Kingdom, France, Russia, Japan and Israel.[41] The use of the Internet by the Falungong movement outside China to spread its message inside the country and around the world has also been met with what looks like a systematic campaign of cyber warfare against its Websites.

Military–industrial nexus

The civilian authorities also have a major role to play in coordinating the mobilisation of computer expertise among the population at large. In part, this means changing the way that people think about information technology by instilling in users a sense of responsibility that will encourage them to install and develop the right kinds of systems for maintaining security. The development of professional support structures is also recommended, in the shape of enterprises dedicated to computer security which can act as 'Internet police' and 'Internet clinics', while also strengthening the research and development into core technologies.[42]

The civilian authorities are also charged with bringing about the improved coordination of the relevant organisations and laws that have developed alongside the growth of the Internet. As Wacker has shown above (Chapter 3), the various agencies concerned with information security have already put in place a comprehensive set of regulations to control domestic activity on the Internet. It is also worth stressing, though, that the development of this regulatory and organisational framework has been largely in response to developments that have taken place outside China. The first arrest that took place under the new raft of regulations was related to international activity, namely the case of Lin Hai. Lin was charged on 25 March 1998 with 'inciting subversion of state power' by providing large numbers of Chinese e-mail addresses to 'hostile foreign publications', such as *VIP Reference*, a newsletter compiled by Chinese democracy advocates in Washington and sent to hundreds of computer users in China. When Wang Youcai was arrested in July 1998 he too was accused of sending e-mail messages to dissidents in the United States while trying to organise an opposition party.

Apart from some efforts to combat domestic computer crime (particularly bank fraud) that began in the early 1980s, the fact that the regulatory project really began in March 1994 can also be seen in part to be a response to international events. This, after all, was the same time that Al Gore announced the Global Information Infrastructure initiative, and just when the Internet was beginning to break out of what Giese calls its 'academic ghetto' in China. Moreover, the reason why the Internet was able to spread beyond the campus at this time was the impetus provided by commercialisation following the lifting of the ban on commercial activity on the Internet by the National Science Foundation of the United States in 1992. It was this policy more than any other that allowed the Internet and the World Wide Web to develop into the popular means of global communication that we know today. Combined with developments in other ICTs such as satellite television, Western policy-makers were increasingly upbeat about the potential power of ICTs to bring about the transformation of authoritarian states. The atmosphere at the time was encapsulated by Rupert Murdoch's famous hailing of satellite television as a threat to totalitarian regimes everywhere, as well as Gore's proclamation that the GII would be a force for the promotion of freedom and democracy.[43]

The security organs in China were thus well aware that there was a widely held belief in the West that the emerging communications networks could be used to exert 'soft power'. As well as developing regulations to control activity, they began to build defensive measures into the architecture of the Internet, such as the restriction of international links to four gateways located in the cities of Beijing, Shanghai and Guangzhou. Regulations introduced in January 2000 require all computer information systems involving state secrets to be neither directly nor indirectly linked with the international Internet.[44] At the same time it was also reported in the PRC press that Chinese companies were forbidden to buy products with encryption software designed by foreign countries, and no domestic organisation or individual would be allowed to sell foreign commercial encryption products.[45]

Policy-makers are aware, however, that such measures taken to prepare for information warfare will remain weak unless China's indigenous technological base can be raised to international standards. As Minister of Information Industry, Wu Jichuan, points out, China must not only adopt the right domestic countermeasures to stand up for its own interests and those of the developing world and avoid becoming an 'information colonial state' or 'semi-colonial state', it must also learn to work with other states. Needless to say, the very rapid pace of technological development and the digital gap with the United States makes this a daunting task. Part of the solution is sought in the integration of the military and civilian information industry sectors.

This kind of integration can be said to have begun, in fact, when military science and technology began to be transferred to the civilian sector in 1985. Integrating the two sectors was further boosted when the Ninth Five-Year Plan (1996–2000) aimed to raise the efficiency and international competitiveness of scientific and industrial research by exposing it to market demand and developing partnerships with foreign firms. Jiang Zemin took the civilian–military link a step further in his report to the Fifteenth Congress of the CPC in September 1997 when he advocated the establishment of an 'orbital defence industry mechanism that interacts with the socialist market economy system'.[46] Section 24 of the current Tenth Five-Year Plan also sees defence industries as being of strategic economic importance. It urges that they should be combined with the civilian sector to promote the task of 'strengthening the armed forces through science and technology', and promises to accelerate the building of 'a technology-intensive army, streamline the armed forces in a Chinese way, increase their capability of fighting defensive wars under conditions of modern technology, especially high technology, and be prepared to meet any contingency'.[47] Some military commentators are also arguing for the armed forces to strengthen logistical management techniques by learning from and helping to strengthen the civilian e-commerce sector, apparently influenced by an initiative in this area taken by the United States Department of Defense in 1999.[48]

The PLA can, in fact, claim to have played a key role in meeting the challenge of the Information Revolution, by reportedly having devoted more than 400 million work days and organised 25 million vehicle trips to participate in and support 10,000-odd key national and local infrastructure projects, including the laying of 20,000 kilometres of optical cable telecommunication lines. It is also claimed that the military has used its advanced scientific and technological achievements over the past five years to support more than 1,000 national economic construction projects, solve urgent problems for more than 150 scientific research projects, transfer some 10,000 scientific and technological findings to the civilian sector, train nearly one million scientific and technological personnel, and help civilian enterprises complete 900-odd technical transformation projects which enabled 320 enterprises to get out of the red and become profitable.[49]

On the civilian side, many of today's key corporations in the information industry sector began to flourish under initiatives such as the '863 Plan', launched in March 1986 as a response to the Reagan administration's 'Strategic Defense Initiative' (or 'Star Wars'), under which the Chinese government aims to promote world-leading high technology firms. In 1999 the State Ministry of Science and Technology decided to make defence-related information technology a high priority within this scheme, bringing academic and scientific research organisations together with large enterprises to lay the foundations for the 'leapfrog-style development' of a new state information infrastructure based on indigenously developed Chinese technology, with a special emphasis on key Internet technologies such as routers.[50] This scheme was given a new shot in the arm in February 2001, when the government marked its fifteenth anniversary with an injection of USD 1.8 bn into the State High-Technology Research and Development Plan, with development of the information industry and especially information security at the top of the agenda.[51] The Ministry of Science and Technology has also established special production bases in Chengdu, Southwest China, and in Shanghai to concentrate on the development and manufacture of security-related information technology.[52]

Some of the key players in the 'national team' of very large enterprise groups that have been fostered by such initiatives to survive in the global market are to be found in the IT sector. The Legend group is a good example. Founded in 1984 with a USD 24,000 loan, by 1999 it had grown to be the largest electronics goods producer in China and the fifth largest in Asia, with its main product being PCs. A merger with the Computing Institute of the Chinese Academy of Sciences (CAS), and financing that derives largely from the Bank of China, gives Legend close links with the state.[53] A similar example is the Capital Iron and Steel Group which announced in March 2001 that it was teaming up with the Beijing Association of Science and Technology to establish an international information automation research centre to engage mainly in intelligence information processing as well as high-tech research and development in complicated system and intelligence control. The automation research institute products they have developed include a dialogue system between humans and machines, advanced robot-controlled machines and lie detectors. Military analysts claim that there have already been encouraging signs in the indigenous development of applied technology and materials technology that can be used to build an information security umbrella. In 2000, the construction of a routing device that can withstand test attacks was hailed as a great success in building a 'strategic pass for information resources entering and leaving national territory and borders'.[54]

Despite such optimism, however, there is also a realisation in China that the preservation of national, economic and personal security is being made ever more complex by the increasingly widespread use of ICTs among the general population. This was one of the points stressed by the director of the Bureau of Public Information Network Security Control (under the Ministry of Public Security), Li Zhao, when he addressed a special meeting on how to deal with the 'Code Red II' virus in August 2001.[55] Yet when it comes to controlling the behaviour of the population, there is already something of a comic air surrounding attempts to stop 'spiritual pollution' through crude measures such as the mass closure of Internet cafes or campaigns by the Beijing municipal authorities to confiscate satellite television receiving equipment.[56] ICTs have in fact already developed beyond the stage where such measures can be effective. As Walton explains, the sheer volume of data that is now flowing across ICTs, fuelled by the move towards broadband, means that the technology used to control communications is moving away from old-style firewalls in favour of dispersing monitoring and censorship architecture throughout the system, down to the level of individual PC platforms.[57] The only way to achieve such levels of sophistication in China is to harness foreign know-how to the cause of strengthening national security.

Western knowledge to preserve Chinese essence

Sometimes foreign know-how can be appropriated directly, by acquiring the information technology necessary for waging information warfare from leading North American and European firms. According to Taiwan's Ministry of Defence, the PRC has introduced advanced technology from Britain and France for use in simulated wars. Walton details how leading North American and European firms take part in annual 'Security China' trade exhibitions and supply crucial assistance for converting the Internet into a massive surveillance system, known as the 'Golden Shield'. Leading foreign firms, he explains, are lured by lucrative contracts with central and local government into helping with the construction of a 'massive, ubiquitous architecture of surveillance', the ultimate aim of which is 'to integrate a gigantic online database with an all-encompassing surveillance network'. This will include linking up cutting-edge technologies such as speech and face recognition, closed-circuit television, smart cards, credit records and Internet surveillance technologies.[58]

Appropriating foreign technology to safeguard national security poses a serious problem, however, because the United States sees maintaining the global dominance of its own information industry as constituting a national security objective.[59] From this perspective, the granting of limited access to the PRC telecommunications and Internet market that was included in the China–United States agreement on PRC accession to the WTO is highly significant as a way by which to introduce foreign technology to China. The trick for the Chinese side was to secure agreement that foreign firms and investors can only operate in the Chinese market if they form partnerships with indigenous firms. Moreover, according to domestic PRC regulations, such partnerships have to be approved by the MII.[60] The MII is thus left with considerable leverage to influence the behaviour of foreign firms in the PRC.

The kind of partnership that is evolving under this formula is already becoming clear as multinational enterprises such as Microsoft, AOL-Time Warner and Hong Kong Telecom form partnerships with members of the PRC's 'national team'. Microsoft paved the way when it struck a deal with Legend in March 1999 to develop boxes to enable Internet access via television sets. At the same time, Microsoft CEO Bill Gates also announced in the Special Economic Zone of Shenzhen a 'strategic cooperation plan' with Rupert Murdoch's Hong Kong Telecom. There is little reason to expect that such firms will operate as agents of 'peaceful evolution' in China. There has been much speculation in the world's press, for example, that Murdoch is only able to play a significant role in the Chinese market because of the considerable lengths to which he has gone to restore his credibility with the Chinese leadership since his 1993 statement about cable television undermining authoritarian regimes. This has included banning the BBC from his Star TV service for China and North Asia, helping Deng Xiaoping's daughter publicise the biography of her father around the world, and criticism of the Dalai Lama for good measure. The *International Herald Tribune* summed up the situation well when AOL-Time Warner chose Legend as its partner to enter the market for Internet services in May 2001, stating: 'Legend enjoys cordial relations with China's regulators and a strong reputation among Chinese consumers – assets that could help offset AOL's lack of operating experience in China and ease apprehensions among Chinese officials and consumers that the company will use its services to download United States culture into China.'[61]

In September 2001 AOL-Time Warner and Murdoch's News Corporation again made significant inroads into the PRC telecoms market when Xu Guangchun, minister of the State Administration of Radio, Film and Television (SARFT), announced that they would be permitted to broadcast directly to a part of Guangdong Province.[62] At the same time, Xu announced that overseas companies (including those listed in Hong Kong and Taiwan) would be forbidden from taking direct equity stakes in mainland cable television concerns, unless they confined themselves to just leasing equipment to local companies. It did not go unnoticed that the way had been paved for the triumphs of AOL and the Murdoch empire through the building of personal links between their top managers and the CCP elite. The head of Star TV, James Murdoch (son of Rupert) is reported to have described the banned Falungong movement as 'dangerous' and an 'apocalyptic cult'. At one dinner in Hong Kong, AOL-Time Warner CEO Gerald Levin is said to have introduced the CCP leader as 'my good friend Jiang Zemin' and 'a man of honour, dedicated to the best interests of his people'.[63] Moreover, it also became apparent that the PRC is not entirely powerless when it comes to spreading its own 'spiritual pollution' around the world when it was revealed that these foreign corporations had agreed to throw their support behind efforts to permit China Central Television's (CCTV) English-language channel to broadcast in the United States.

It is important to note, then, that Chinese policy-makers do not see any incompatibility between maintaining national information security and working in harmony with foreign interests. This also applies to the way in which the Chinese government is trying to work in accordance with the practices of international society. Indeed, a considerable part of the MII-CCP report is concerned with explaining the nature of legislation introduced by technologically advanced states to control the use of ICTs.[64] Jiang Zemin himself has emphasised that the internationalisation of the information network demands regulation at the international level and has called for more active Chinese participation in the organisations that draw up relevant treaties, as well as a stepping up of international exchanges and cooperation in this field.[65]

In this respect, PRC policy-makers are aware that their own efforts to maintain information borders can be considerably strengthened by the growing international awareness that the globalisation of ICTs poses a threat to the security of all sovereign states. In fact, it is precisely because the Internet does not recognise borders that leaving any part of it unregulated will create a loophole for activities that can destabilise any other part of the world. Such activities range from organised crime syndicates launching large-scale attacks against financial systems, to drug dealing, money laundering, the spreading of child pornography and terrorism.[66]

Just as in the field of conventional warfare, states have already been made painfully aware that mutual restraint is needed to avoid unneccessary mutual damage being inflicted by information warfare. Internationally, the possibility of new kinds of destruction has already led to a growing movement to modernise the laws of war to accommodate the information age. In particular, such a development implies the evolution of a new interpretation of the UN Charter and customary international law that can accommodate the definition of cyber warfare as the use of force. Without such a definition, it will be difficult to decide what constitutes legitimate self-defence. Moreover, when such definitions are decided, they will have to be made enforceable by the construction of multilateral treaties that facilitate tracking, attribution and transnational enforcement.[67]

There are already indications that electronic information warfare is starting to conform with this dynamic anyway. When Taiwan's hackers responded to assaults on their island's computer systems from the PRC in 1999 with eight waves of their own attacks, for example, the chaos became so great on both sides that calls for a ceasefire went out. Similarly, the PRC authorities were made painfully aware of the consequences of encouraging cyber warfare to be launched from their own territory after hacking attacks against United States targets following the Hainan spy-plane incident triggered off counter-attacks by American hackers. An official of the State Office for Computer Network and Information Security claimed that 13.8 per cent of attacks on international networks between the middle of April and early May that year had been aimed at mainland China. One technician claimed that the networks of the enterprise he worked for had been probed and scanned no less than 80,000 times a day, with 100 actual attacks per day.[68]

The disincentives for engaging in information warfare become even more serious when other governments see the PRC as a likely adversary in any digital conflict. The United States is clearly the most significant potential adversary in this respect. But technologically advanced societies like Japan and Taiwan are also very important. In August 1999, for example, the Ministry of National Defence of the ROC on Taiwan announced that it had established a committee to deal with information warfare to counter moves by the mainland, which would invite experts and party representatives to study its comprehensive strategy to combat information warfare.[69] Among twelve measures announced in the National Defense Policy White Paper issued by the Democratic Progressive Party of Taiwan, just before Chen Shui-bian won the presidential election in March 2000, was the deployment of digital forces and the development of corresponding doctrines to increase the flexibility, mobility, and general readiness of the island's standing forces.[70] Such plans have not gone without notice in Mainland China, especially that for the

ICTs and international security 153

establishment of a special Internet warfare unit, the 'Tiger Brigade' (*laohu dui*), in January 2000, as well as the talent that the island has shown for creating computer viruses.[71]

The need for self-restraint and regulation at the international level becomes even more pressing when third-party states are liable to be drawn into cyber conflicts, either due to the way in which packets of data travel through their part of the Internet as they find the least congested route between any two points in the world, or through the deliberate use by hackers of servers in third countries to launch attacks on their enemies. In the wake of the Hainan spy-plane incident, for example, South Korea's Ministry of Information and Communication felt the need to warn government organisations, universities and private institutions to take precautions against Chinese and American hackers using their Internet sites as a stopover to attack each other's computer systems. Seoul also set up a special task force under the Korea Information Security Agency (KISA) to provide professional support and advice for possible victims.[72]

Although an international legal structure for controlling information warfare is still lacking, however, conditions within which cooperation on international information security can take place are already being put in place. One aspect of this is a steady convergence between the domestic legislation enacted within various states around the world. Sometimes the parallels are striking. For example, Chinese legislation now requires ISPs to keep records of all content and all users that appear on their servers for scrutiny by the security agencies if required. In the United Kingdom, meantime, the Regulation of Investigatory Powers (RIP) Act also requires ISPs to retain all communications data originating or terminating in the United Kingdom, or routed through United Kingdom networks. Employers in the United Kingdom are permitted to monitor the e-mail of their staff, and the Home Office is considering granting powers to the security agencies to have access to records of every phone call, e-mail and internet connection made in Britain. The director general of the national criminal intelligence service, Roger Gaspar, even compared the proposed new data bank to the national DNA database under development.[73]

In addition to this convergence between the domestic regulations introduced by states around the world, well before the events of 11 September 2001 it was also clear that interconnectivity was driving states with very different political systems and cultures to move towards international collaboration on issues of information security. In November 2000, for example, a network was cracked that involved the use of the Internet by criminals in China and the Republic of China on Taiwan to illicitly siphon off money from a South African bank.[74] Such successful police action must have resulted from extensive cooperation between the security agencies from both sides of the Taiwan Strait, yet their governments do not even talk to each other.

Collaboration between states can also be seen in the growing tendency to share information on individuals and organisations that is accumulated on digital databases, again well before 11 September 2001. When the United Kingdom's House of Lords held an inquiry into this phenomenon in 1999, it was so concerned that it felt the need to issue a strong warning about the dangers of giving in to pressures from third-party countries for access to data on EU databases while it remained unclear which data protection rules could be applied and which body, if any, was responsible for supervising data flows.[75] Among the states with which the EU has been exploring the possibility of exchanging data accumulated on its various intelligence databases were the United States and Russia.[76] This chain could be extended further by the fact that Russia is a member of the 'Shanghai Six', under which it cooperates with the PRC, Uzbekistan, Kazakhstan, Kyrgyzstan and Tajikistan to maintain security in Central Asia. This is the region, of course, where the PRC has long been engaging in a 'strike hard' campaign against the secessionist movement of the Islamic Uighur population in Xinjiang.

The events of 11 September 2001 have, of course, immensely strengthened this tendency towards international cooperation on issues of state security. The agreement signed on 21 October 2001 at the APEC summit in Shanghai committed several states, among them the United States, the PRC and Russia, to taking measures to counter 'all forms of terrorist acts'. This includes working together to strengthen activities to protect critical sectors, including telecommunications; cooperation to develop electronic movement records systems that will enhance border security; and strengthening capacity building and economic and technical cooperation to enable member economies to put into place and enforce effective counter-terrorism measures.[77] At the global level, too, states have been called on by the UN Security Council to accelerate and intensify the exchange of operational information regarding the actions or movements of terrorist organisations, and specifically in relation to their use of information technology.[78] It needs hardly be stated that the concept of 'terrorism' is yet to be defined by international law. The activities that the Security Council connects it with are broad enough, though, including 'transnational organized crime', trade in illicit drugs, money laundering, illegal arms trafficking and the movement of potentially deadly materials. Perhaps the most visible indication of how the pendulum has swung from notions of justice towards international order is the uncertain fate of the CIA's project to sponsor SafeWeb to provide software to enable anonymity for Internet users in authoritarian states such as the PRC.[79]

Conclusion

It has been argued above that the information revolution and the spread of the Internet is seen to pose a threat to the national security of the PRC by elements in the military, the government, the CCP, academics and the general population. Responses are thus being devised at all levels, ranging from military doctrine and training, to domestic regulation, industrial policy, and international cooperation. Perhaps the range of policy positions being proposed is best encapsulated by the various recommendations made by delegates to the Chinese People's Political Consultative Conference (CPPCC), when this united front organisation discussed issues of network security in March 2001. Yang Yixian, a CPPCC delegate with professional expertise in the development of Internet security systems, suggested that a nongovernmental organ be set up between government departments and enterprises according to international practice, with the task of managing all problems involving network security in a unified way and avoiding possible loopholes caused by the barriers between different state bureaucracies or regions. Hong Kong delegate Lau Nai Keung stressed the need to make full use of his territory's international status to set up an authoritative international cooperative organisation that could intensify international cooperation in the management of the network. Mi Zhenyu of the PLA's Academy of Military Sciences, on the other hand, proposed that China's information industry should concentrate its energies on developing indigenous software and hardware products. Meanwhile, Xu Wenbo, secretary-general of the 'Network Civilization Project Organizing Committee', urged the government to intensify control, examine and screen unhealthy contents, and promote national culture in the network environment.[80]

It is the task of top-level authorities like the MII to render such different recommendations into a coherent whole. The response to the various suggestions made at the CPPCC meeting from Zhang Chunjiang, Vice Minister for Information Industry (one of the authors of the MII-CCP report discussed above), was that building network security is indeed a complex job calling for legal support, technical guidance, and cultural involvement. From the perspective of bureaucratic politics, it might be added, there is also a piece of the cake for just about anybody who can portray the information age as a threat to national security. In this respect, it is important to note that the above discussion took place in the context of the unveiling of the Tenth Five-Year Plan.

Yet, despite the complexities of the security problems generated by digitalisation, it has been argued above that policy-makers also look to the increasing global interconnectivity of digital networks as a source of strength when it comes to building network security. When looking at some of the more extreme visions of information warfare, we would do well to remember that this is not the first Revolution in Military Affairs to have taken place in history. The age-old need for states to exercise self-restraint and engage in cooperation for the sake of maintaining order and security provides reasons for believing that the dangers posed by information warfare will have to be dealt with at the international level in the same way that other kinds of technological developments have eventually fallen under regimes of global governance. Events since 11 September 2001 have served to reinforce this tendency. It is not hard to unravel this paradox when we think about the nature of state security. As Buzan points out: 'States of all types benefit from the widespread feeling among individuals that anything is better than reversion to the state of nature. So long as the state performs its Hobbesian task of keeping chaos at bay, this service will be seen by many to offset the costs of other state purposes, whatever they may be.'[81]

Seen from the longer historical perspective of Chinese nation-building, policy-makers in the PRC thus face the task of harnessing the forces that are generated by economic and technological globalisation in ways that buttress national information security rather than erode it. In many ways, this is an interesting extension of the nineteenth-century neo-Confucian formula of using Western functional knowledge (*yong*) to preserve Chinese essence (*ti*).[82] Or, as Jiang Zemin put it in a July 1991 speech to commemorate the seventieth anniversary of the founding of the CCP, 'take the ancient to serve the modern, the foreign to serve China' (*gu wei jin yong, yang wei zhong yong*).[83] It is thus that China's long revolution continues into the information age.

Notes

1 ICANN, the Internet Corporation for Assigned Names and Numbers, is a private organisation established by the Clinton administration to administer the addresses upon which the direction of digital traffic on the Internet depends. See Chapter 6 by Ermert and Hughes in this volume for the politics and international controversies surrounding this organisation.
2 On the relative strengths of information industry in the PRC and the United States see P. Nolan and M. Hasecic, 'China and the Third Industrial Revolution', in *Cambridge Review of International Affairs*, vo. 13, no. 2, pp. 164–80.
3 Ryan Henry and E.C. Peartree, 'Military Theory and Information Warfare', in Henry and Peartree (eds), *The Information Revolution and International Security*, Washington DC: CSIS Press, 1998, p. 108.
4 H.H. Frederick, *Global Communication and International Relations*, California: Wadsworth, 1993, pp. 220–2.
5 R.O. Keohane and J.S. Nye Jnr, 'Power and Interdependence in the Information Age', *Foreign Affairs*, 1998, vol. 77, no. 5, p. 86.
6 CSIS Taskforce, Global Organized Crime Project, *Cybercrime . . . Cyberterrorism . . . Cyberwarfare . . . Averting an Electronic Waterloo*, Washington: CSIS Press, 1998.
7 J. Arquilla and D. Ronfeldt, 'Information Power and Grand Strategy: In Athena's Camp', in S.J.D. Schwartzstein (ed.), *The Information Revolution and National Security*, Washington DC: CSIS Press, 1996.
8 Henry and Peartree, 'Military Theory and Information Warfare', p. 116.
9 CSIS Taskforce, *Cybercrime*.
10 Henry and Peartree, 'Military Theory and Information Warfare', p. 120.
11 J.A. Baker III, 'America in Asia: Emerging Architecture for a Pacific Community', *Foreign Affairs*, 1991/2, vol. 70, no. 5, p. 16.
12 Al Gore, 'Remarks Prepared for Delivery to International Telecommunications Union, Buenos Aires', 21 March 1994. Online. Available HTTP: <http://www.iitf.nist.gov/documents/speeches/032194_gore_giispeech.html> (accessed 20 February 2002).
13 M. Albright, 'Permanent Normal Trade Relations for China, Remarks at Agilent Technologies Health Care Solutions Group', 6 April 2000. Online. Available HTTP: <http://secretary.state.gov/www/statements/2000/000406.html> (accessed 3 April 2002).

14 M. Albright, 'Address to the World Trade Center', 9 May 2000, Denver, Colorado. Online. Available HTTP: <http://secretary.state.gov/www/statements/2000/000509a.html> (accessed 6 February 2002).
15 Deng Xiaoping, '*Women you xinxin ba zhongguo de shiqing zuo de geng hao*' ('We Have the Confidence to Conduct China's Affairs Better'), *Deng Xiaoping Wenxuan, di san juan* (*Selected Works of Deng Xiaoping, Vol. 3*), Beijing: Renmin chubanshe, 1993, pp. 325–6.
16 Yu Xiaoqiu, '*Dui xinxi jishu yu guojia anquan ruogan wenti sikao*' ('Some Considerations on Information Technology and National Security'), *Xiandai guoji guanxi* (*Contemporary International Relations*), 2001, vol. 3, pp. 6–12.
17 Zhang Chunjiang and Ni Jianmin, *Guojia xinxi anquan baogao* (*Report on National Information Security*), Beijing: Renmin chubanshe, 2000, pp. 37–8.
18 Zhang and Ni, *Guojia xinxi anquan baogao*, pp. 4–5.
19 Zhang and Ni, *Guojia xinxi anquan baogao*, pp. 4–5.
20 Y. Liu and W. Zhang, 'High-Tech Development and State Security', *Jiefangjun bao* (*PLA News*), 11 January 2000, p. 6. English version in BBC Summary of World Broadcasts, FE/3764 G/6, 15 February 2000.
21 See the special edition on information warfare of 'PLA Military Affairs Salon Online' ('*Jiefangjun bao wang ban jun shi shalong*') of *Jiefangjun bao* (*PLA News*). Available HTTP: <http://www.pladaily.com.cn/item/vote/hacker/content/002.htm> (accessed 14 May 2002), and the special edition on 'The Logistics Front' ('*houqin zhanxian*'). Online. Available HTTP: <http://www.pladaily.com.cn> (accessed 14 May 2002).
22 See in particular, Song Shikui '*Jianli kuaqu lianhe baozhang wangluo*' ('Establish a Cross Region United and Secure Internet'), *Guofang bao*, 12 July 2001, p. 3. Online. Available HTTP: <http://www.pladaily.com.cn/gb/jskj/2001/07/12/20010712017055_jslt.html> (accessed 14 May 2002); Yang Liuguo, '*Zhishi houqin baozhang de te zheng*' ('Characteristics of Securing Knowledge Logistics'), *Guofang bao*, 4 June 2001, p. 3. Online. Available HTTP: <http://www.pladaily.com.cn/gb/hqzx/2001/06/04/20010604017060_hqggrd.html> (accessed 14 May 2002); Weng Changling, '*Xinxi fangwei – xinxi zhanzheng de zhongyao yi huan*' ('Information Defense – An Important Link in Information Warfare'). Online. Available HTTP: <http://www.pladaily.com.cn/item/vote/houqing/content/7-015.htm> (accessed 14 May 2002); and the various comments in the special edition on logistics in *Jiefangjun bao*, 10 April 2001, p. 6. Online. Available HTTP: <http://www.pladaily.com.cn/gb/hqzx/2001/04/10/200104100001081_hqggrd.html> (accessed 14 May 2002).
23 '*Wo jun weisheng xinxihua jianshe chuju guimo*' ('Informatisation of the Military Medical System Takes Shape'), *Jiefangjun bao* (*PLA News*), 6 November 2001, p. 1. Online. *Jiefangjun bao wangluo ban–jun shi shalong*. Available HTTP: <http://www.pladaily.com.cn/gb/hqzx/2001/11/06/20011106001006_hqggrd.html> (accessed 14 May 2002).
24 Department of Defense (United States), *Quadrennial Defense Review Report, September 20 2001*. Online. Available HTTP: <http://www.defenselink.mil/pubs/qdr2001.pdf> (accessed 14 May 2002). For a Chinese military commentary on the Department of Defense report, see Teng Fei, '*21 shiji Meiguo jundui fazhan de qushi*' ('Trends of United States Military Development in the 21st Century'). Online. Available HTTP: <http://www.pladaily.com.cn/item/vote/houqing/content/7-002.htm> (accessed 14 May 2002).
25 Yang Shisong and Wu Hao, '*Zhuizong wang shang duxiao*' ('In Search of the Online Poison Peddlars'), *Jiefangjun bao wangluo ban–jun shi shalong*. Online. Available HTTP: <http://www.pladaily.com.cn/item/vote/hacker/content/008.htm> (accessed 14 May 2002).

26 Ma Yaxi, 'Gouzhu "wangluo changcheng"' ('Construct an "Internet Great Wall"'). Online. Available HTTP: <http://www.pladaily.com.cn/item/vote/hacker/content/002.htm> (accessed 14 May 2002).
27 Liu and Zhang, 'High-Tech Development and State Security'.
28 'General Fu Quanyou calls for accelerating army's modernization' BBC Monitoring, Global Newsline, Asia Pacific Political File, 10 April 2001. Originally from *Xinhua News Agency*, Beijing (domestic service, in Chinese), 10 April 2001.
29 For overviews of changes in Chinese strategic thinking since the 1980s see N. Li, 'The PLA's Evolving Warfighting Doctrine, Strategy and Tactics, 1985–95: A Chinese Perspective', *China Quarterly*, 1996, no. 146, pp. 443–63; P.H.B. Godwin, 'From Continent to Periphery: PLA Doctrine, Strategy and Capabilities Towards 2000', in the same volume, pp. 464–87.
30 Dai Yifang, 'Amplify the Guidance Role of Military Theories, Ensure the Smooth Implementation of the "Two Basic Changes"', *Guofang*, 1997, vol. 5, pp. 4–5.
31 Jiang Zemin, '*Gaoju Deng Xiaoping lilun weida qizhi, ba jianshe you Zhongguo tese shehui zhuyi shiye quanmian tuixiang ershiyi shiji*' ('Hold High the Great Banner of Deng Xiaoping Theory, Take the Task of Building Socialism With Chinese Characteristics Forwards to the 21st Century'), in *Zhonggong shiwu da baogao budao duben* (*Guide to the CCP 15th Party Congress*), Hong Kong: Mingliu chubanshe, 1997, p. 32.
32 V. Lai, 'ROC Defense Ministry Sets Up Information Warfare Committee', Asia Intelligence Wire, Central News Agency, 16 August 1999.
33 *Jiefangjun bao*, March 24, 1998, cited in CSIS Taskforce, *Cybercrime*, p. xvi.
34 Henry and Peartree, 'Military Theory and Information Warfare', p. 122.
35 'Air Force Publishes First Information Warfare Teaching Aid', BBC Monitoring, Global Newsline, Asia Pacific Political File, 20 April 2001. Original Chinese version in *Zhongguo Xinwen She* (China News Society), Beijing, 20 April 2001.
36 'Japanese Newsletter on Taiwan Information Defense Concerns'. Online. Available HTTP: <http://www.usembassy-china.org.cn/english/sandt/taiinfowar.html> (accessed 14 November 2002).
37 Qiao Liang and Wang Xianghui, *Chaoxian zhan: dui quanqiuhua shidai zhanzheng yu zhanfa de xiangding* (*Unlimited War: Doctrine for War and Tactics in the Age of Globalization*), Beijing: Jiefangjun wenyi chubanshe, 1999, pp. 141–58.
38 Qiao and Wang, *Chaoxian zhan*, p. 44.
39 Lee Teng-hui first proposed his two states theory on 9 July 1999 in response to questions submitted by *Deutsche Welle* (*Voice of Germany*). The figure of 7,200 attacks was given to a meeting of legislators by Zhang Guangyuan, head of the Information Office of the ROC National Security Bureau. *Lianhe bao* (*United Daily News*) (overseas edition), 17 August 1999, p. 3.
40 '*Hulianwang shang de jiaoliang*' ('Showdown on the Internet'), *Beijing Qingnian Bao* (*Beijing Youth News*), 11 May 1995, p. 11.
41 Qian Fang, '*Fangwei "heike": junshi shang de yi ge jipo renwu*' ('Protect Against "Hackers": An Urgent Task in Military Affairs'), *Jiefangjun bao wangluo ban – jun shi shalong* (*PLA Daily Online – Military Affairs Salon*). Online. Available HTTP: <http://www.peopledaily.com.cn/item/vote/hacker/content/txtyindaqianghacker.htm> (accessed 14 May 2002).
42 Wu Jichuan, '*Gouzhu mianxiang 21 shiji de guojia xinxi anquan tixi*' ('Construct a National Information Security System to Face the 21st Century'), Preface to Zhang and Ni, *Guojia xinxi anquan baogao*, pp. ii–iv.
43 Al Gore, 'Remarks Prepared for Delivery to International Telecommunications Union'.

44 Public Security Bureau, 'Provisions on Secrecy Management of Computer Information Systems on the Internet', promulgated 1 January 2000. English translation available online. Available HTTP: <http://www.usembassy-china.org.cn/english/sandt/netsecret.htm> (accessed 15 October 2001).
45 *Yangcheng wanbao*, Guangzhou, 20 Feb 2000. Online. Reprinted in SWB FE/3772 G/9, 24 February 2000. Original HTTP not provided.
46 Jiang Zemin, '*Gaoju Deng Xiaoping lilun weida qizhi*', p. 32.
47 '*Shiwu jihua gangyao quanwen*' (*Complete Text of the Outline of the Tenth Five Year Plan*). Online. Available HTTP: <http://www.chinaemb.or.kr/chn/9272.html> (accessed 10 April 2002).
48 Liao Rugeng, '*Miandui dianzi houqin, women que shenme?*' ('Facing Electronic Logistics, What do We Lack?'). Online. Available HTTP: <http://www.pladaily.com.cn/gb/jskj/2001/01/31/20010131002004_jslt.html> (accessed 14 May 2002).
49 State Council, *White Paper – China's National Defense*, Information Office of the State Council (PRC), 1998. Online. Available HTTP: <http://tigger.uic.edu/~rodrigo/white_paper_98.htm> (accessed 5 April 2002).
50 'China Develops Router Technology for High-Speed Internet Use', *Xinhua News Agency*, Beijing (domestic service, Chinese), 9 August 2001. English version in BBC Monitoring, Global Newsline, Asia Pacific Economic file, 21 September 2001.
51 'China to Invest 15 Billion Yuan in High-tech Development', *Xinhua News Agency*, Beijing (English), 14 February 2001.
52 'China Sets Up Information Security Production Base in Southwest', *Xinhua News Agency*, Beijing (English), 17 August 2001.
53 On the 'national team' see D. Sutherland, 'Policies to Build National Champions: China's "National Team" of Enterprise Groups', in P. Nolan (ed.), *China and the Global Business Revolution*, Basingstoke and New York: Palgrave, 2001, pp. 67–140.
54 Liu and Zhang, 'High-Tech Development and State Security'.
55 'China to Boost Efforts Against Internet, Network Related Crimes', *Xinhua News Agency*, Beijing (domestic service, Chinese), 27 August 2001. English version in BBC Monitoring, Global Newsline, Asia Pacific Political File, 29 August 2001.
56 'China to Crack Down on Illegal Satellite TV Receiving Facilities', *Xinhua News Agency*, Beijing (domestic service, Chinese), 12 December 2001. English version in BBC Monitoring, Global Newsline, Asia Pacific Political File, 14 December 2001.
57 G. Walton, *China's Golden Shield: Corporations and the Development of Surveillance Technology in the People's Republic of China*, Montreal: International Centre for Human Rights and Democratic Development, 2001. Online. Available HTTP: <http://www.ichrdd.ca/frame.iphtml?langue=0> (accessed 29 October 2001).
58 Walton, *China's Golden Shield*.
59 See Department of Defense (US), *Quadrennial Defense Review Report*. See also CSIS Taskforce, *Cybercrime . . . Cyberterrorism . . .*, p. 66. For an historical overview of the development of the Clinton administration's policies towards ICTs see Ethan Kapstein, *Hegemony Wired: American Politics and the New Economy*, Paris: Institut des relations internationales, 2000.
60 State Council, '*Hulianwang xinxi fuwu guanli banfa*', ('Methods for Managing Internet Information Service'), 25 September 2000. Online. Available HTTP: <http://www.cnnic.net.cn/policy/18.shtml> (accessed 9 December 2001).

61 Clay Chandler, 'AOL Picks Partner for China Foray', *International Herald Tribune*, 5 June 2001, p. 13.
62 'Tune Into China', *Financial Times*, 5 September 2001, p. 16.
63 D. Gittings and J. Borger, 'Homer and Bart Realise Murdoch's Dream of China Coup', *The Guardian*, 6 September 2001, p. 6.
64 Zhang and Ni, *Guojia xinxi anquan baogao*, pp. 271–84.
65 'China: President Urges Tighter Controls, More Political Debate on Internet', *Xinhua News Agency*, Beijing (domestic service, Chinese), 11 July 2001. English version in BBC Monitoring, Global Newsline, Asia Pacific Political File, 13 July 2001.
66 B. Schneier, *Secrets and Lies: Digital Security in a Networked World*, New York: John Wiley and Sons Inc., 2000, pp. 21, 67.
67 G.D. Grove, S.E. Goodman and S.J. Lukasik, 'Cyber Attacks and International Law', *Survival*, 2000, vol. 42 no. 3, pp. 89–104.
68 'Chinese Computer Security Official Interviewed on Recent Hacking', original in *Xinhua News Agency*, Beijing (domestic service, Chinese) 3 May 2001. English version in BBC Monitoring, Global Newsline, Asia Pacific Political File, 4 May 2001.
69 Lai, 'ROC Defense Ministry Sets Up Information Warfare Committee'.
70 Taiwan Democratic Progressive Party, *National Defense Policy White Paper*, 1999. Online. Available HTTP: <http://www.taiwandc.org/dpp-pol2.htm> (accessed 7 April 2002).
71 Wang Jun, *'Taiwan zhuzhong wangluo zhan'* ('Taiwan Pays Attention to Netwar'), *Jiefangjun bao wangluo ban – jun shi shalong (PLA Daily Online – Military Affairs Salon)*. Online. Available HTTP: <http://www.pladaily.com.cn/item/vote/hacker/content/010.htm> (accessed 14 May 2002).
72 *Yonhap News Agency*, Seoul, in English, 6 May 2001.
73 R. Norton-Taylor, 'Spies Seek Access to Phone E-Mail and Net Links', *The Guardian*, 4 December 2000, p. 8.
74 *'Nanfei daohui an, Chen Jiyang tou an'* ('South African Financial Scandal, Chen Jiyang Involved), *Lianhe bao (United Daily News)*, (overseas edition), 11 November 2000, p. 3.
75 House of Lords, Select Commitee on the European Communities, *European Union Databases*, (23rd Report, Session 1998–99), London: The Stationery Office, 1999.
76 House of Lords, Select Committee on the European Communities, *European Union Databases*, p. 12.
77 APEC, 'APEC Leaders Statement on Counter-Terrorism', 21 October, 2001. Online. Available HTTP: <www.apecsec.org.sg> (accessed 10 April 2002).
78 UN Security Council, 'Resolution 1373', 28 September 2001. Online. Available HTTP: <http://www.un.org/Docs/scres/2001/res1373e.pdf> (accessed 12 April 2002).
79 On SafeWeb, see Chapter 3 by Gudrun Wacker in this volume.
80 'CPPCC Members Discuss Issue of Network Security', *Xinhua News Agency* for Hong Kong, Beijing, 1 March 2001.
81 Buzan, Barry, *People States and Fear* (Second Edition), New York and London: Harvester Wheatsheaf, 1991, p. 43.
82 The *ti–yong* formula of saving Chinese cultural essence (*ti*) by using Western functional knowledge (*yong*) was formulated by the Confucian reformer Zhang Zhidong (1837–1909). See Joseph Levenson, *Confucian China and Its Modern Fate*, Berkeley and Los Angeles: University of California, 1965, pp. 59–79.
83 Jiang Zemin, *'Jianshe you Zhongguo tese de shehui zhuyi wenhua'* ('Build a Socialist Culture with Chinese Characteristics'), in *Mao Zedong, Deng*

Xiaoping, Jiang Zemin lun shijie guan rensheng guan jiazhi guan (*Mao Zedong, Deng Xiaoping and Jiang Zemin on Concepts of World View, Humanity and Values*), Central Party Documentation Research Dept of CCP (ed.), Hong Kong: Mingliu chubanshe, 1998, p. 380.

Bibliography

Newspapers and agencies consulted

BBC Summary of World Broadcasts (SWB)
Beijing Evening News (Beijing wanbao)
Beijing Review
Beijing Youth News (Beijing qingnian bao)
Caijin Shibao (Financial Times)
The Economist
Far Eastern Economic Review
Financial Times
The Guardian
Guofang (National Defence)
International Herald Tribune
Ming Pao (Hong Kong)
People's Daily (Remin ribao)
People's Liberation Army News (Jiefangjun bao)
Sing Tao Jih Pao (Hong Kong)
South China Morning Post
Taipei Times
United Daily News (Lianhe bao), (Taiwan: overseas edition)
Xinhua News Agency
Yonhap News Agency (Seoul)

Books and articles

Albright, M., 'Address to the World Trade Center', 9 May 2000, Denver, Colorado. Online. Available HTTP: <http://secretary.state.gov/www/statements/2000/000509a.html> (accessed 6 February 2002).
—— 'Permanent Normal Trade Relations for China, Remarks at Agilent Technologies Health Care Solutions Group', 6 April 2000. Online. Available HTTP: <http://secretary.state.gov/www/statements/2000/000406.html> (accessed 3 April 2002).
APEC, 'APEC Leaders Statement on Counter-Terrorism', 21 October, 2001. Online. Available HTTP: <www.apecsec.org.sg> (accessed 19 April 2002).
APEC China Secretariat, 'e-APEC Strategy Published', 22 October 2001. Online. Available HTTP <http://www.apec-china.org.cn/APEC2001/20011022/928262.htm> (accessed 9 November 2001).

—— 'Shanghai Accord'. Online. Available HTTP: <http://www.apec-china.org.cn/APEC2001/20011021/927941.htm> (accessed 9 November 2001).
Arquilla, J. and Ronfeldt, D., 'Information Power and Grand Strategy: In Athena's Camp', in S.J.D. Schwartzstein (ed.), *The Information Revolution and National Security*, Washington DC: CSIS Press, 1996.
Baker III, J.A., 'America in Asia: Emerging Architecture for a Pacific Community', *Foreign Affairs*, 1991/2, vol. 70, no. 5, pp. 1–18.
Boas, T. and Kalathil, S., 'The Internet and State Control in Authoritarian Regimes: China, Cuba, and the Counterrevolution'. Carnegie Working Paper no. 21, July 2001. Online. Available HTTP: <http://www.ceip.org/files/pdf/21KalathilBoas.pdf> (accessed 10 May 2002).
Boyle, J., 'Foucault in Cyberspace: Surveillance, Sovereignty, and Hard-Wired Censors', 1997. Online. Available HTTP: <http://www.wcl.american.edu/pub/faculty/boyle/foucault.htm> (accessed 6 November 2000).
Buruma, I., 'China in Cyberspace', *New York Review of Books*, 4 November 1999, vol. 46, no. 17, pp. 9–12.
Buzan, B., *People States and Fear* (Second Edition), New York and London: Harvester Wheatsheaf, 1991.
Caro, R.A., *The Power Broker: Robert Moses and the Fall of New York*, New York: Random House, 1974.
Cha, Ariana Eunjung, 'Bye-Bye Borderless Web: Countries Are Raising Electronic Fences', *International Herald Tribune*, 5 January 2002, pp. 1, 4.
Chandler, A.D. Jr, *The Visible Hand: The Managerial Revolution in American Business*, Cambridge, MA: Belknap, Harvard University Press, 1977.
Chen, Gary, 'China's Booming Internet Sector: Open or Closed To Foreign Investment?', 8 October 1999. Online. Available HTTP: <http://www.chinaonline.com/industry/infotech/NewsArchive/Secure/1999/october/C9100519REV-SS.asp> (accessed 9 January 2000).
Chen, Judy M., 'Willing Partners to Repression?', 27 November 2000, *Digital Freedom Network*. Online. Available HTTP: <http://www.dfn.org/focus/china/multinationals.htm> (accessed 28 February 2001).
Chen, Shui-bian, 'President Chen Shui-bian's Inauguration Speech', 20 May 2000. Online. Available HTTP: <http:// members.tripod.com/Ken_Davies/inaugural.html> (accessed 27 February 2002).
—— 'The Third Way for Taiwan: A New Political Perspective', 6 December 1999. Online. Available HTTP: <http://www.president.gov.tw/1_president/e_subject-04a.html> (accessed 27 January 2000).
Chen, Xiaoning, '*Lun youxian yu xinmeiti de guanxi*' ('On the Relationship Between Cable TV and New Media'), parts 1 to 5. Online. Available HTTP: <http://www.sarft.com> (accessed 2 February 2001).
Chen, Yichong, '*Zhongguo chuanmei gu huo zai ziben yu zhengce jiafen zhong*' ('The Stock of China's Media is Struggling Between Capital and Policies'). Online. Available HTTP: <http://tech.sina.comcn/ite/75784.shtml> (accessed 25 July 2001).
China Internet Information Centre 'How Many Users Are There in China', 8 February 2001. Online. Available HTTP <http://www.chinaguide.org/english/7235.htm> (accessed 29 March 2001).
ChinaOnline, 'China's Internet Development Timeline'. Online. Available HTTP: <http://www.chinaonline.com/issues/internet_policy/c9101571.asp> (accessed 10 November 2000).

CNNIC (China Internet Network Information Centre), 'Evolution of the Internet in China'. Online. Available HTTP: <http://www.cnnic.net.cn/evolution.shtml> (accessed 9 April 2002).
—— *Zhongguo hulianwangluo daikuan baogao* (*China Internet Bandwidth Report*), November 2001. Online. Available HTTP: <http://www.cnnic.net.cn/mapinfo/rep2001-11/2.shtml> (accessed 18 January 2002).
—— *Zhongguo Internet fazhan zhuangkuang tongji baogao (1997/10) (China Internet Development Statistics Report (1997/10))*. Online. Available HTTP: <http://www.cnnic.net.cn/develst/report1.shtml> (accessed 4 May 2001).
—— *Zhongguo Internet fazhan zhuangkuang tongji baogao (1998/7) (China Internet Development Statistics Report (1998/7))*. Online. Available HTTP: <http://www.cnnic.net.cn/develst/report2.shtml> (accessed 4 May 2001).
—— *Zhongguo Internet fazhan zhuangkuang tongji baogao (1999/1) (China Internet Development Statistics Report (1999/1))*. Online. Available HTTP: <http://www.cnnic.net.cn/develst/report3.shtml> (accessed 4 May 2001).
—— *Zhongguo Internet fazhan zhuangkuang tongji baogao (1999/7) (China Internet Development Statistics Report (1999/7))*. Online. Available HTTP: <http://www.cnnic.net.cn/develst/report4.shtml> (accessed 4 May 2001).
—— *Zhongguo Internet fazhan zhuangkuang tongji baogao (2000/1) (China Internet Development Statistics Report (2000/1))*. Online. Available HTTP: <http://www.cnnic.net.cn/develst/cnnic2000.shtml> (accessed 4 May 2001).
—— *Zhongguo Internet fazhan zhuangkuang tongji baogao (2000/7) (China Internet Development Statistics Report (2000/7))*. Online. Available HTTP: <http://www.cnnic.net.cn/develst/cnnic200007.shtml> (accessed 4 May 2001).
—— *Zhongguo Internet fazhan zhuangkuang tongji baogao (2001/1) (China Internet Development Statistics Report (2001/1))*. Online. Available HTTP: <http://www.cnnic.net.cn/develst/cnnic200101.shtml> (accessed 4 May 2001).
—— *Zhongguo Internet fazhan zhuangkuang tongji baogao (2001/7) (China Internet Development Statistics Report (2001/7))*. Online. Available HTTP: <http://www.cnnic.net.cn/develst/rep200107-1.shtml> (accessed 19 July 2001).
—— *Zhongguo hulianwangluo fazhan zhuangkuang tongji baogao (2002/1) (China Internet Development Statistics Report (2002/1))*. Online. Available HTTP: <http://www.cnnic.net.cn/develst/2002-1/> (accessed 25 January 2002).
Cowhig, D., 'New Net rules Not a Nuisance?', 5 December 2000. Online. Available HTTP: <http://www.chinaonline.com/commentary_analysis/internet/NewsArchive/secure/2000/December/c00120160.asp> (accessed 6 December 2000).
CSIS Taskforce, Global Organized Crime Project, *Cybercrime . . . Cyberterrorism . . . Cyberwarfare . . . Averting an Electronic Waterloo*, Washington DC: CSIS Press, 1998.
Dai, Xiudian, 'Chinese Politics of the Internet: Control and Anti-control', *Cambridge Review of International Affairs*, 2000, vol. 13, no. 2, pp. 181–94.
—— *The Digital Revolution and Governance*, Aldershot, Ashgate, 2000.
—— 'Towards a Digital Economy with Chinese Characteristics?', paper presented at 'Development and Impact of the Internet in China' workshop, London School of Economics, 8–10 December 2000.
Dai, Yifang, 'Amplify the Guidance Role of Military Theories, Ensure the Smooth Implementation of the "Two Basic Changes"', *Guofang*, 1997, vol. 5, pp. 4–5.

Dean, Ten, 'Telecommunications: The Data Communications Market Opens Up'. Online. Available HTTP: <http://www.chinaonline.com> (accessed 28 December 2001).

Democratic Progressive Party (DPP), Taiwan, *National Defense Policy White Paper*, (1999). Online. Available HTTP: <http://www.taiwandc.org/dpp-pol2.htm> (accessed 7 April 2002).

Deng Shoupeng, 'Development Status and Prospects of E-Commerce'. Online. Available HTTP: <http://www.telecomn.com/english/china_comm> (accessed 15 November 2000).

Deng, Xiaoping, '*Women you xinxin ba zhongguo de shiqing zuo de geng hao*' ('We have the Confidence to Conduct China's Affairs Better'), *Deng Xiaoping Wenxuan, di san juan*, (*Selected Works of Deng Xiaoping, Vol. 3*), Beijing: Renmin chubanshe, 1993, pp. 325–6.

Denton, T., 'The Governance of the Domain Name System'. Online. Available HTTP: <http://www.tmdenton.com/Speeches/The%20Governance%20of%20 the%20Domain%20Name%20System.htm> (accessed 5 April 2002).

Department of Defense (United States), *Quadrennial Defense Review Report, September 20 2001*. Online. Available HTTP: <http://www.defenselink.mil/pubs/qdr2001.pdf> (accessed 14 May 2002).

Department of Foreign Affairs and International Trade, Canada, 'The China Business Collection: China's Western Development Strategy', *China Perspectives 2001*. Online. Available HTTP: <http://www.dfait-maeci.gc.ca/china/business/WesternDev-e.asp> (accessed 29 August 2001).

Dordick, H. and Wang, G., *The Information Society: A Retrospective View*, London: Sage, 1993.

Drake, W.J., Kalathil, S. and Boas, T.C., 'Dictatorships in the Digital Age: Some Considerations on the Internet in China and Cuba', *iMP*, October 2000. Online. Available HTTP: <http://www.cisp.org/imp/october_2000/10_00drake.htm> (accessed 11 November 2000)

Eckblad, J.G. 'Let them Eat Web. The Internet: Viagra or Sugar Pill for Treating Fledgling Economies?', *Student Technology Forum*. Online. Available HTTP: <http://www.ncsu.edu/connect/josh1.html> (accessed 25 April 2001).

European Commission, *The Information Society and Development: A Review of the EC's Experience in Asia, Latin America and the Mediterranean*, DG External Relations, ER/04 Economic Analysis, Brussels, 12 January 2001.

—— *The Information Society and Development: The Role of the European Union*, Communication to the Council and the European Parliament, COM(97) 351 final, 15 July 1997. Online. Available HTTP: <http://europa.eu.int/ISPO/intcoop/i_com_97_351.html> (accessed 15 February 2002).

Executive Yuan, Council for Economic Planning and Development, 'Plan to Develop Knowledge-based Development in Taiwan', Taipei: Council for Economic Planning and Development, September 2000.

—— National Science Council, 'White Paper on Science and Technology', Taipei: National Science Council, 2 December 1997. Online. Available HTTP: <http://www.stic.gov.tw/policy/scimeeting/E-whitepaper/summary_e.html> (accessed 27 July 2001).

—— 'Action Plan for Building a Technologically Advanced Nation', Taipei: National Science Council, April 1998.

Fackler, M. 'The Great Fire Wall of China?', 8 November 2000. Online. Available

HTTP: <http://www.abcnews.go.com/sections/tech/DailyNews/chinanet001108.html> (accessed 28 February 2001).

Fischer, D., 'Rückzug des Staates aus dem chinesischen Mediensektor? Neue institutionelle Arrangements am Beispiel des Zeitungsmarktes', paper presented at the first meeting of ASC (Arbeitskreis sozialwissenschaftliche Chinaforschung) 'Funktionswandel des Staates', 17–18 November 2000' Witten.

Forney, M. 'Taipei's Tech-Talent Exodus', 21 May 2001. Online. Available HTTP: <http://www.time.com/time/asia/news/printout/0,9788,109642,00.html> (accessed 5 December 2001).

Foucault, M., *Discipline and Punish. The Birth of the Prison*, London: Penguin, 1991.

Franda, M., *Launching into Cyberspace. Internet Development and Politics in Five World Regions*, London: Lynne Riener, 2002.

Fravel, T., 'The Bureaucrats' Battle over the Internet in China', 17 February 2000. Online. Available HTTP: <http://www.virtualchina.com/news/feb00/021800-ministries-tf.html> (accessed 19 February 2000).

Frederick, H.H., *Global Communication and International Relations*, California: Wadsworth, 1993.

G8, 'Okinawa Charter on Global Information Society', Okinawa, 22 July 2000. Online. Available HTTP <http://www.library.utoronto.ca/g7/summit/2000okinawa/gis.htm> (accessed 13 October 2000).

Gebler, D. 'Chinese Web Filter May Block Western Sites'. Online. Available HTTP: <http://www.newsfactor.com/perl/printer/7805/> (accessed 28 February 2001).

Giese, K., 'Big Brother mit rechtstaatlichem Anspruch. Gesetzliche Einschränkungen des Internet in der VR China', in B. Engels and O. Nielinger (eds), *Elektronischer Handel in Afrika, Asien, Lateinamerika und Nahost*, Hamburg: Deutsches Überseeinstitut, pp. 127–52 (Schriften des Deutschen Überseeinstituts, no. 50).

—— 'China und die APEC', *China aktuell*, October 2001, pp. 1087–100.

—— 'Das gesetzliche Korsett für das Internet ist eng geschnürt', *China aktuell*, October 2000, pp. 1173–81.

—— 'Government Online: A Very Partial Success', 10 October 1999. Online. Available HTTP: <http://www.chinabiz.org/articles/it/991005.htm> (accessed 20 December 2000).

—— 'Internet, E-Business und Digital Divide in der VR China. Eine kritische Bestandsaufnahme', *China aktuell*, January 2001, pp. 33–47.

Godwin, P.H.B., 'From Continent to Periphery: PLA Doctrine, Strategy and Capabilities Towards 2000', *China Quarterly*, 1996, no. 146, pp. 464–87.

Gore, A., 'Remarks Prepared for Delivery to International Telecommunications Union, Buenos Aires', 21 March 1994. Online. Available HTTP: <http://www.iitf.nist.gov/documents/speeches/032194_gore_giispeech.html> (accessed 20 February 2002).

Grove, G.D., Goodman S.E. and Lukasik, S.J. 'Cyber Attacks and International Law', *Survival*, 2000, vol. 42 no. 3, pp. 89–104.

Guo Liang and Bu Wei, '*2000 nian Beijing, Shanghai, Guangzhou, Chengdu, Changsha Qingshao nian hulianwang shiyong zhuangkuang ji yingxiang de tiaocha baogao*' (*Survey Report on Internet Usage and Influence in Beijing, Shanghai, Guangzhou, Chengdu and Changsha in the Year 2000*), April 2001. Online. Available HTTP: <http://www.chinace.org/ce/itre/index_.htm> (accessed 9 August 2001).

Hachigian, N. 'China and the Net: A Love–Hate Relationship'. Online. Available HTTP: <http://www.chinaonline.com/commentary/archive/secure/2011/March/c01030260.asp> (accessed 8 March 2001).

Han, Xia, '*Zhashi tuijin woguo de kuandai jianshe he yingyong*' (Forging the Construction and Application of China's Broadband), 23 August 2001. Online. Available HTTP: <http://www.mii.gov.cn> (accessed 20 December 2001).

Hartford, K., 'Cyberspace With Chinese Characteristics', *Current History*, September 2000, vol. 99, no. 638, pp. 255–62. Online. Available HTTP: <http://www.pollycyber.com/pubs/ch/home.htm> (accessed 10 June 2001).

Heng, Toh-mun and Low, L. (eds), *Regional Cooperation and Growth Triangles in ASEAN*, Singapore: Times Academic Press, 1995.

Henry, Ryan and Peartree, E.C., 'Military Theory and Information Warfare', in Henry and Peartree (eds), *The Information Revolution and International Security*, Washington DC: CSIS Press, 1998, pp. 105–27.

Holbig, H. 'Reformanlauf ins neue Jahrhundert – Offizielle und inoffizielle Agenda der 5. Plenartagung des XV. ZK', *China aktuell*, October 2000, pp. 1167–72.

Hong Kong Democratic Foundation, 'Policy Paper: Response to 1999/2000 Budget', 5 December 1999. Online. Available HTTP: <http://www.hkdf.org/papers/990512budget.htm> (accessed 6 December 2000).

House of Lords, Select Committee on the European Communities, *European Union Databases*, (23rd Report, Session 1998–99), London: The Stationery Office, 1999.

Hsu, Jinn-yuh and Saxenian, A., 'The Limits of Guanxi Capitalism: Transnational Collaboration between Taiwan and the USA', *Environment and Planning*, 2000, vol. 32, 11, pp. 1991–2005.

Hughes, C.R., 'Nationalism in Chinese Cyberspace', *Cambridge Review of International Affairs*, Spring/Summer 2000, vol. 13, no. 2, pp. 195–209.

IETF (Internet Engineering Task Force), 'The Tao of the IETF: A Novice's Guide (RFC 3160)'. Online. Available HTTP: http://www.ietf.org/tao.html#intro (accessed 21 May 2002).

Janda, R., 'Benchmarking A Chinese Offer on Telecommunications: Context and Comparisons', *International Journal of Communications Law and Policy*, 1999, issue 3. Online. Available HTTP <http://www.ijclp.org> (accessed 14 February 2000).

Jessop, B. and Sum, Ngai-Ling, 'An Entrepreneurial City in Action: Hong Kong's Emerging Strategies in and for (Inter-)Urban Competition', *Urban Studies* (Special Issue on Asia's Global Cities), 2000, vol. 33, no. 3, pp. 2287–313.

Jia, Hepeng, 'Getting its Foot in China Door', 30 October 2001. Online. Available HTTP: <http://www.chinadaily.net/bw/2001-10-30/41368.html> (accessed 14 December 2001).

Jiang, Zemin, '*Gaoju Deng Xiaoping lilun weida qizhi, ba jianshe you Zhongguo tese shehui zhuyi shiye quanmian tuixiang ershiyi shiji*' ('Hold High the Great Banner of Deng Xiaoping Theory, Take the Task of Building Socialism With Chinese Characteristics Forwards to the 21st Century'), in *Zhonggong shiwu da baogao budao duben* (*Guide to the CCP 15th Party Congress*), Hong Kong: Mingliu chubanshe, 1997, pp. 1–43.

—— '*Jianshe you Zhongguo tese de shehui zhuyi wenhua*' ('Build a Socialist Culture with Chinese Characteristics'), in *Mao Zedong, Deng Xiaoping, Jiang Zemin lun shijie guan rensheng guan jiazhi guan* (*Mao Zedong, Deng Xiaoping and Jiang*

Zemin on Concepts of World View, Humanity and Values), Central Party Documentation Research Dept of CCP (ed.), Hong Kong: Mingliu chubanshe, 1998, pp. 379–80.

Jordan, A. and Khanna, J., 'Economic Interdependence and Challenges to the Nation-State: the Emergence of Natural Economic Territories in the Asia-Pacific', *Journal of International Affairs*, 1995, vol. 48, no. 2, pp. 433–62.

Kapstein, E., *Hegemony Wired: American Politics and the New Economy*, Paris: Institut des relations internationales, 2000.

Keohane, R.O. and Nye Jnr, J.S., 'Power and Interdependence in the Information Age', *Foreign Affairs*, 1998, vol. 77, no. 5, pp. 80–94.

Kluver, A.R., 'New Media and the End of Nationalism: China and the US in a War of Words', *Mots Pluriels*, August 2001, no. 18. Online. Available HTTP: <http://www.arts.uwa.edu.au/MotsPluriels/MP1801ak.html> (accessed 29 August 2001).

Kristof, N.D., 'The Chip on China's Shoulder', *New York Times*, 18 January 2002. Online. Available HTTP: <http://www.nytimes.com/2002/01/18/opinion/18KRIS.html> (accessed 18 January 2002).

Lai, V., 'ROC Defense Ministry Sets Up Information Warfare Committee', Asia Intelligence Wire, Central News Agency, 16 August 1999.

Lam, Willy Wo-lap, 'Hi-Tech Confucian Future' in 'Analysis', *South China Morning Post*, 13 October 1999, p. 8.

Lee, Chong-moon, Miller, W., Hancock, M. and Rowen, H. (eds), *The Silicon Valley Edge: A Habitat for Innovation and Entrepreneurship*, Stanford, CA: Stanford University Press, 2000.

Lessig, L., 'Who's Holding Back Broadband?', *Washington Post*, 9 January 2002. Online. Available HTTP: <http://www.newsbytes.com/news/02/173496.html> (accessed 10 January 2002).

—— *Code and Other Laws of Cyberspace*, New York: Basic Books, 1999.

Levenson, J., *Confucian China and Its Modern Fate*, Berkeley and Los Angeles: University of California, 1965.

Li, N., 'The PLA's Evolving Warfighting Doctrine, Strategy and Tactics, 1985–95: A Chinese Perspective', *China Quarterly*, 1996, no. 146, pp. 443–63.

Liang, Ping, '*Tuijin woguo sanwang ronghe jincheng de sikao*' ('Considerations about How to Accelerate the Process of China's Tri-Net-Convergence'), *Hulianwang shijie (Netcom World)* 2002, no. 4, pp. 29–31.

Liao, Rugeng, '*Miandui dianzi houqin, women que shenme?*' ('Facing Electronic Logistics, What do We Lack?). Online. Available HTTP: <http://www.pladaily.com.cn/gb/jskj/2001/01/31/20010131002004_jslt.html> (accessed 14 May 2002).

Liu, Sunray, 'China Plans to Leapfrog Western Digital TV Specs', *EE Times*, 26 January 2001.Online. Available HTTP: <http://www.eet.com/story/OEG20010126S0032> (accessed 29 December 2001).

—— 'China Stands at Broadband's Gate', *EE Times*, 23 April 2001. Online. Available HTTP: <http://www.eetimes.com/story/OEG20010423S0122> (accessed 29 December 2001).

—— 'China To Run Broadband Network Over Cable-TV Lines', *EE Times*, 13 May 1999. Online. Available HTTP: <http://content.techweb.com/wire/story/TWB19990513S0006> (accessed 21 February 2000).

Liu, Y. and Zhang, W., 'High-Tech Development and State Security', *Jiefangjun bao (PLA News)*, 11 January 2000, p. 6. English version in BBC Summary of World Broadcasts, FE/3764 G/6, 15 February 2000.

LoBaido, A.C., 'Life with Beijing's Bruisers'. Online. Available HTTP: <http://www.worldnetdaily.com/news/printer-friendly.asp?ARTICLE_ID=21446> (accessed 28 February 2001).

Ma, Yaxi, *'Gouzhu "wangluo changcheng"'* ('Construct an "Internet Great Wall"'). Online. Available HTTP: <http://www.pladaily.com.cn/item/vote/hacker/content/002.htm> (accessed 14 May 2002).

Ma Zhongshi, 'Business Goes West; Regional Growth is a Major National Priority', *CHINA 2000*. Online. Available HTTP: <http://www.china2thou.com/003p2.htm> (accessed 29 August 2001).

Marshall, T. and Kuhn, A., 'China Goes One-on-One With the Net', *LA Times*, 27 January 2001.Online. Available HTTP: <http://www.latimes.com/business/cutting/lat_chitek010127.htm> (accessed 29 January 2001).

McGill, D.C., 'Sina.com's Delicate Balancing Act', 23 May 2000. Online. Available HTTP: <http://www.virtualchina.com/finance/stirfry/052300-stirfry-dcm-alo2.html> (accessed 25 May 2000).

Ministry of Information Industry (MII), *2001 nian 12 yue zhuyao tongxin zhibiao zhaiyaobiao* (*December 2001 Summary of Major Statistical Indicators for Telecommunications*). Online. Available HTTP <http://www.mii.gov.cn/mii/hyzw/tongji/yb/tongjiziliao200112.htm> (accessed 7 February 2002).

—— *'Dianxin yewu jingying xukezheng guanli banfa'* ('Managment Methods for Licences in the Telecommunication Business'), 1 January 2002. Online. Available HTTP: <http://www.mii.gov.cn/mii/zcfg/bl19.htm> (accessed 28 February 2002).

—— and SARFT, *'Guanyu jiaqiang guangbo dianshi youxian wangluo jianshe guanli de yijian'* ('Opinions on Strengthening the Management of the Construction of the Broadcasting and TV Cable Network'), 13 September 1999, *Zhonghua Renmin Gongheguo Guowuyuan Gongbao*, 1999, no. 35, pp. 1573–5, document 82 approved by the State Council Office, 1999.

—— *'Hulianwang dianzi gonggao fuwu guanli guiding'* ('Provisions for the Administration of Electronic Information Services on the Internet'), 8 October 2000, *Guowuyuan gongbao* 2001, no. 2, pp. 45–6.

—— *'Zhonghua renmin gongheguo dianxin tiaoli'* ('Telecommunication Regulations of the People's Republic of China'), 25 September 2000, *Guowuyuan gongbao* 2000, no. 33, pp. 11–21. Online. Available HTTP: <http://www.mii.gov.cn/news2000/1013_1.htm> (accessed 2 December 2000).

Mueller, M. and Tan, Zixiang, *China in the Information Age: Telecommunications and the Dilemmas of Reform*, Westport, CT and London: Praeger Publishers, 1997.

National Bureau of Statistics, *China Statistical Yearbook 2000*, Beijing: China Statistics Press, 2000 (CD-ROM edition).

—— *China Statistical Yearbook 2001*, Beijing: China Statistics Press, 2001 (CD-ROM edition).

National People's Congress Standing Committee, *'Guanyu weihu hulianwang anquan de jueding'* ('Resolution of the Standing Committee of the National People's Congress on Maintaining Security of Computer Networks'), 28 December 2000, *Guowuyuan gongbao*, 2001, no. 5, pp. 21–3.

Naughton, J., *A Brief History of the Future: The Origins of the Internet*, London: Weidenfeld and Nicolson, 1999.

Neumann, A.L., 'The Great Firewall', *CPJ Briefings: Press Freedom Reports*.

Online. Available HTTP: <http://www.cpj.org/Briefings/2001/China_jan01/China_jan01.html> (accessed 23 February 2001).
Nolan, P. and Hasecic, M., 'China and the Third Industrial Revolution', *Cambridge Review of International Affairs*, vol. 13, no. 2, pp. 164–80.
OECD, *Is There a New Economy? First Report on the OECD Growth Project*. Paris: OECD, June 2000.
—— *Special Issue on Information Infrastructure*, STI Review, no. 20, Paris: OECD, 1997.
Perkmann, M. and Sum, Ngai-ling (eds) *Globalization, Regionalization and Cross-Border Regions*, Basingstoke: Palgrave, 2002.
Persaud, A., 'The Knowledge Gap', *Foreign Affairs*, March/April 2001, pp. 107–17.
Plafker, T. 'Tapping China's Potential. Logistical Barriers Slow Push for Online Learning', *International Herald Tribune*, 16 October 2000. Online Available HTTP: <http://62.172.206.162/IHT/SR/101600/sr101600k.html> (accessed 10 January 2001).
Public Security Bureau, 'Provisions on Secrecy Management of Computer Information Systems on the Internet', promulgated 1 January 2000. English translation available online. Available HTTP: <http://www.usembassy-china.org.cn/english/sandt/netsecret.htm> (accessed 15 October 2001).
Pye, L.W., *The Mandarin and the Cadre: China's Political Cultures*, Ann Arbor, Michigan: Center for Chinese Studies, University of Michigan, 1988.
Qian, Fang, 'Fangwei "heike": junshi shang de yi ge jipo renwu' ('Protect Against "Hackers": An Urgent Task in Military Affairs'), *Jiefangjun bao wangluo ban – jun shi shalong (PLA Daily Online – Military Affairs Salon)*. Online. Available HTTP: <http://www.peopledaily.com.cn/item/vote/hacker/content/txtyindaqianghacker.htm> (accessed 14 May 2002).
Qiao, Liang and Wang, Xianghui, *Chaoxian zhan: dui quanqiuhua shidai zhanzheng yu zhanfa de xiangding (Unlimited War: Doctrine for War and Tactics in the Age of Globalization)*, Beijing: Jiefangjun wenyi chubanshe, 1999.
Qiu, Jack Linchuan, 'Chinese Opinions Collide Online', *Online Journalism Review*. Online. Available HTTP: <http://ojr.usc.edu/content/story.cfm?request=561> (accessed 17 April 2001).
—— 'Internet Censorship in China (1999–2000)', *Communications Law in Transition Newsletter*, 18 February 2001, vol. 2, no. 3. Online. Available HTTP: <http://pcmlp.socleg.ox.ac.uk/transition/issue2_3/qiu.htm> (accessed 25 February 2002).
—— 'Virtual Censorship in China: Keeping the Gate between the Cyberspaces', *International Journal of Communications Law and Policy*, Winter 1999/2000, issue 4. Online. Available HTTP: <http://111.ijclp.org/4_2000/ijclp_webdoc_1_4_2000.html> (accessed 10 November 2000), pp. 1–25.
Robinson, N., 'New Laws Seek to Balance Privacy and Surveillance', *Jane's Intelligence Review*, January 2002, vol. 14, no. 1, pp. 52–3.
Rodan, G., 'The Internet and Political Control in Singapore', *Political Science Quarterly*, 1998, vol. 113, no. 1, pp. 63–89.
Rosenberg, D., *Cloning Silicon Valley*, London: Pearson Education, 2002.
SARFT (State Administration of Radio, Film and Television). Online. Available HTTP: <www.sarft.com/site/termp/sho.htm?temp=T201112345&d1=20011203 838093> (accessed 2 December 2001).

Saxenian, A., 'Networks of Immigrant Entrepreneurs' in Lee, Chong-moon, Miller, W., Hancock, M. and Rowen, H. (eds), *The Silicon Valley Edge*, pp. 248–75.
—— *Regional Advantage: Culture and Competition in Silicon Valley and Route 128*, Cambridge, MA: Harvard University Press, 1994.
Schneier, B., *Secrets and Lies: Digital Security in a Networked World*, New York: John Wiley and Sons Inc., 2000.
Scott, B.R., 'The Great Divide in the Global Village', *Foreign Affairs*, January/February 2001, pp. 160–77.
Shapiro, A.L., 'The Internet', *Foreign Policy*, Summer 1999, no. 115, pp. 14–27.
Sieren, F., 'Von Netzen und Mauern. Über die Substanz chinesischer Internet-phantasien', in K. Leggewie and C. Maar (eds), *Internet & Politik. Von der Zuschauer- zur Beteiligungsdemokratie*, Cologne: Bollmann, 1998, pp. 229–35.
Song, Shikui *'Jianli kuaqu lianhe baozhang wangluo'* ('Establish a Cross Region United and Secure Internet'), *Guofang bao*, 12 July 2001, p. 3. Online. Available HTTP: <http://www.pladaily.com.cn/gb/jskj/2001/07/12/20010712017055_jslt.html> (accessed 14 May 2002).
State Council, *White Paper – China's National Defense*, Information Office of the State Council (PRC), 1998. Online. Available HTTP: <http://tigger.uic.edu/~rodrigo/white_paper_98.htm> (accessed 5 April 2002).
—— *'Chuban guanli tiaoli'* ('Regulations Governing the Administration of the Publishing Industry'), 1 January 1997, *Guowuyuan gongbao*, 1997, no. 2, pp. 38–46.
—— *'Chuban guanli tiaoli'* ('Regulations Governing the Administration of the Publishing Industry'), 25 December 2001, *Guowuyuan gongbao*, 2002, no. 4, pp. 14–20.
—— *'Guowuyuan Bangongting guanyu chengli Guojia xinxihua gongzuo lingdao xiaozu de tongzhi'* ('Circular of the State Council Office on establishing the Leading Group for Informatization'), 23 December 1999, *Guowuyuan gongbao*, 2000, no. 6, pp. 8–9
—— *'Hulianwang xinxi fuwu guanli banfa'* ('Methods for the Administration of Internet-Based Information Services'), 20 September 2000, *Guowuyuan gongbao*, 2000, no. 34, pp. 7–9. Online. Available HTTP: <http://www.peopledaily.com.cn/GB/channel5/28/200010017/2557566.html> (accessed 24 October 2000).
—— *'Hulianwang xinxi fuwu guanli banfa'*, ('Methods for Managing Internet Information Service'), 25 September 2000. Online. Available HTTP: <http://www.cnnic.net.cn/policy/18.shtml> (accessed 9 December 2001).
—— *'Shangyong mima guanli tiaoli'* ('Regulations for the Administration of Commercial Encryption'), State Council Directive No. 273, December 1999, *Guowuyuan gongbao*, 1999, no. 36, pp. 1,663–7.
—— *Shiwu jihua gangyao quanwen* (*Complete Text of the Outline of the Tenth Five Year Plan*). Online. Available HTTP: <http://www.chinaemb.or.kr/chn/9272.html> (accessed 10 April 2002).
—— *'Hulianwangzhan congshi dengzai xinwen yewu guanli zhanxing guiding'* (Interim Provisions for the Administration of Release of News by Websites), 6 November 2000, *Guowuyuan gongbao*, 2001, no. 2, pp. 46–8.
State Secrets Bureau, *'Zhonghua Renmin Gongheguo baoshou guojia mimi fa shishi banfa'* ('Provisions Governing the Implementation of the State Secrets Law of the People's Republic of China'), *Guowuyuan gongbao*, 1990, no. 14, pp. 538–43.

Sum, Ngai-Ling, 'Time–Space Embeddedness and Geo-Governance of Cross-Border Regional Modes of Growth: their Nature and Dynamics in East Asian Cases', in A. Amin and J. Hausner (eds), *Beyond Market and Hierarchy*, Cheltenham: Edward Elgar, 1997, pp. 159–95.

—— 'Globalization and Hong Kong's Entrepreneurial City Strategies: Contested Visions and the Remaking of City Governance in (Post-)Crisis Hong Kong', in J. Logan (ed.), *The New Chinese City: Globalization and Market Reform*, Oxford: Blackwell, 2002, pp. 74–91.

—— 'Rethinking Globalization: Re-articulating the Spatial Scale and Temporal Horizons of Trans-Border Spaces', in K. Olds, P. Dicken, F. Kelly, L. Kong and H. Yeung (eds), *Globalization and the Asia-Pacific: Contested Territories*, London: Routledge, 1999, pp. 129–46.

Sussman, G., *Communication, Technology and Politics in the Information Age*. London: Sage, 1997.

Sutherland, D., 'Policies to Build National Champions: China's "National Team" of Enterprise Groups', in P. Nolan (ed.), *China and the Global Business Revolution*, Basingstoke and New York: Palgrave, 2001, pp. 67–140.

Tan, Tin Wee, 'Policy and Coordination Issues in Multilingual Internet Names'. Online. Available HTTP: http://www.itu.int/mdns/presentations/dayone/tan1.ppt (Microsoft Powerpoint presentation), (accessed 30 May 2002).

Tan, Zixiang, Foster, W. and Goodman, S., 'China's State-Coordinated Internet Infrastructure', *Communications of the ACM*, June 1999, vol. 42, no. 6 , pp. 44–52.

Tao, David, 'Asia Cable TV – A Broadband Service and Content Provider in PRC'. Online. Available HTTP: <http://www.asiacabletv.com> (accessed 2 January 2001).

Tao, Wenzhao, 'Censorship and Protest: The Regulation of BBS in China People Daily', *first monday*. Online. Available HTTP: <http://www.firstmonday.dk/issues/issue6_1/tao/index.html> (accessed 23 February 2001).

Teng, Fei, '*21 shiji Meiguo jundui fazhan de qushi*' ('Trends of United States Military Development in the 21st Century'). Online. Available HTTP: <http://www.pladaily.com.cn/item/vote/houqing/content/7-002.htm> (accessed 14 May 2002).

Thant, Myo and Tang, Min and H. Kakazu (eds), *Growth Triangles in Asia: A New Approach to Regional Economic Cooperation*, Hong Kong: Oxford University Press, 1998.

Thomas, S., 'Das Internet in China. Teil 1: Aufbau einter Informkationsinfrastruktur, *China aktuell*, May 1999, pp. 500–10.

—— 'Das Internet in der VR China. Teil 2: Nutzung und Inhalte von Online-Medien', *China aktuell*, June 1999, pp. 596–606.

Toffler, A., *The Third Wave*. London: Pan Books, 1981.

Tsang, D. Yam-keung, 'Financial Secretary's Transcript on Cyberport', Hong Kong: Hong Kong SAR Government, 17 March 1999. Online. Available HTTP: <http://www.info.gov.hk/gia/general/199903/17/0317146.htm> (accessed 6 December 1999).

—— 'Onward with New Strength', budget speech delivered by the Financial Secretary in the Legislative Council Meeting, Hong Kong: Hong Kong SAR Government, 3 March 1999.

Tung, Chee-Hwa, 'From Adversity to Opportunity', policy speech delivered by the Chief Executive in the Legislative Council Meeting, Hong Kong: Hong Kong SAR Government, 7 October 1998.

United Nations Development Programme (UNDP), *Human Development Report 2001: Making New Technologies Work for Human Development.* Oxford: Oxford University Press, 2001.

UNDP in China, 'Project Brief – CPR/00/202', 19 February 2001. Online. Available HTTP: <http://unchina.org/undp/news/html/010220-1.html> (accessed 4 April 2001).

—— 'UNDP Launched Its Project of "Poverty Reduction through Access to Information, Communication and Technologies" in China', 20 February 2001. Online. Available HTTP: <http://unchina.org/undp/news/html/010220.htm> (accessed 4 April 2001).

Wacker, G., 'Chinesische Reaktionen auf die Terroranschläge in den USA'. Online. Available HTTP: <http://www.swp-berlin.org/produkte/brennpunkte/wnd11sep 6C.htm> (accessed 21 November 2001).

—— 'Widerstand ist zwecklos: Internet und Zensur in China', in G. Schucher (ed.), *Asien und das Internet*, Hamburg: Institut für Asienkunde, pp. 70–96.

Walton, G., *China's Golden Shield: Corporations and the Development of Surveillance Technology in the People's Republic of China*, Montreal: International Centre for Human Rights and Democratic Development, 2001. Online. Available HTTP: <http://www.ichrdd.ca/frame.iphtml?langue=0> (accessed 29 October 2001).

Wang, Jun, '*Taiwan zhuzhong wangluo zhan*' ('Taiwan Pays Attention to Netwar'), *Jiefangjun bao wanluo ban – jun shi shalong* (*PLA Daily Online – Military Affairs Salon*). Online. Available HTTP: <http://www.pladialy.com.cn/item/vote/hacker/content/010.htm> (accessed 14 May 2002).

Wang Qiming, 'Opportunities and Challenges. A Case of Internet Development in China', last updated 8 February 2001. Online. Available HTTP: <http://www.dse.de/ef/digital/wang-e.htm> (accessed 17 April 2001).

Wang Xiangdong, 'Mobile Communication and Mobile Internet in China'. Online. Available HTTP: <http://www.telecomvisions.com/articles/pdf/china_mobile_internet.pdf> (accessed 7 February 2002).

Webb, D., 'Cyber Villas by Sea', 22 March 1999. Online. Available HTTP: <http://www.webb-site.com/articles/cybervillas.htm> (accessed 12 December 1999).

Weng, Changling, '*Xinxi fangwei – xinxi zhanzheng de zhongyao yi huan*' ('Information Defense – An Important Link in Information Warfare'). Online. Available HTTP: <http://www.pladaily.com.cn/item/vote/houqing/content/7-015.htm> (accessed 14 May 2002.

Winkel, O., 'Sicherheit in der digitalen Informationsgesellschaft', *Aus Politik und Zeitgeschichte*, 6 October 2000, no. B 41–42, pp. 19–30.

Winner, L., 'Do Artifacts Have Politics?' in Donald Mackenzie and Judy Wajcman, *The Social Shaping of Technology* (Second Edition), Buckingham and Philadelphia: Open University Press, 1999.

World Bank, *China's Development Strategy: The Knowledge and Innovation Perspective*. Washington DC: World Bank, 2000.

—— *Knowledge for Development*. Washington DC: World Bank, 1998/99.

Wu, Jichuan, '*Gouzhu mianxiang 21 shiji de guojia xinxi anquan tixi*' ('Construct a National Information Security System to Face the 21st Century'), Preface to Zhang and Ni, *Guojia xinxi anquan baogao*, 2000.

Xiao, Hongzhi, '*Dianxin fuwu qushi tansuo – dalu dianxun shichang xiankuan pingxi*' ('Probing the Trend of the Telecommunications Sector – A Critical

Assessment of the Telecommunications Market in Mainland China'), 8 May 2001. Online. Available HTTP: <http://www.find.org.tw/0105/trend/0105_trend_disp.asp?trend_id=1158> (accessed 3 December 2001).
Xu, Ante and Armstrong, P., *Chinese Telecom Market. A Study Report by Northern Business Information*. New York: McGraw-Hill Companies, 1995.
Yang, Dali L., 'The Great Net of China'. Online. Available HTTP: <http://www.mfcinsight.com/article/010209/oped4.html> (accessed 16 February 2001).
Yang, Haifeng, '*Zhuanjia lun xiayidai wangluo*' ('Experts on the Next Generation of the Internet'), *Huliangwang Shijie*, 2002, no. 1, pp. 61–2.
Yang, Liuguo, '*Zhishi houqin baozhang de te zheng*' ('Characteristics of Securing Knowledge Logistics'), *Guofang bao*, 4 June 2001, p. 3. Online. Available HTTP: <http://www.pladaily.com.cn/gb/hqzx/2001/06/04/20010604017060_hqggrd.html> (accessed 14 May 2002).
Yang, Shisong and Wu, Hao, '*Zhuizong wang shang duxiao*' ('In Search of the Online Poison Peddlars'), *Jiefangjun bao wang ban jun shi shalong*. Online. Available HTTP: <http://www.pladaily.com.cn/item/vote/hacker/content/008.htm> (accessed 14 May 2002).
Yang, Wei, '*Shuzi dianshi de fazhan ji gei xiangguan chanye dailai de jiyu*' ('Development of Digital TV and Its Impact on Related Industries'). Online. Available HTTP: <http://www.chinactv.com/jslw/ztbg/85.htm> (accessed 27 December 2001).
Yao, Yin, '*Kuandainian weihe weichen qihou*' ('Why There was No Atmosphere of "The Year of Broadband"'), *Tongxin Xinxibao*, 14 November 2001, p. 3.
Yu, Xiaoqiu, '*Dui xinxi jishu yu guojia anquan ruogan wenti sikao*' ('Some Considerations on Information Technology and National Security'), *Xiandai guoji guanxi* (*Contemporary International Relations*), 2001, vol. 3, pp. 6–12.
Zhang, Chunjiang and Ni, Jianmin, *Guojia xinxi anquan baogao* (*Report on National Information Security*), Beijing: Renmin chubanshe, 2000.
Zhang Junhua, 'China's "Government Online" and Attempts to Gain Technical Legitimacy', *Asien*, July 2001, no. 80, pp. 93–115.
—— 'Chinas steiniger Weg zum E-Government', in G. Schucher (ed.), *Asien und das Internet*, Hamburg: Institut für Asienkunde, 2002, pp. 97–111.
—— 'Will the Government "Serve the People"? The Development of Chinese E-government', *New Media and Society*, vol. 4, no. 2, June 2002, pp. 163–84.
Zhao Jin and Xu Hongzhou, *Xinxi Zhongguo* (*Information China*), Beijing: Jingji kexue chubanshe, 2001.
Zhen, Yong, '*Sanwang ronghe caidian deshi, kuandai qudong dianzi shangwu*' ('Colour-TV Will Benefit from "Tri-Network Convergence", Broadband Will Forge E-commerce via TV'), *Beijing wanbao*, 6 April 2001, p. 5.
Zhou, Qiren, *Shuwang jingzheng (Competition Among Telecommunication Networks)*, Beijing: Sanlian shudian, 2001.
Zhu, Jonathan J.H. and Zhou He, 'Information Accessibility, User Sophistication, and Source Credibility: The Impact of the Internet on Value Orientations in Mainland China', *Journal of Computer-Mediated Communication*, January 2002, vol. 7, no. 2. Online. Available HTTP: <http://www.ascusc.org/jcmc/vol7/issue2/china.html> (accessed 21 February 2002).
Zita, K., 'LMDS and Broadband Local Networks for Asia', 9 July 1999. Online. Available HTTP: <http://www.vii.org/papers/ptc97.htm> (accessed 30 December 2001).

Index

863 Program 9, 148

Acer: Digital Services Corp. 116; Group 108, 109, 116, 117; Net 116; Software Capital Group 116
agriculture 8, 21, 50
Albright, Madeleine 140
AOL 60; Time-Warner 134, 150–1
Asia Pacific Economic Cooperation (APEC) 38, 66; Shanghai 154
Asian financial crisis 5, 85, 103–4

Baker, James 140
Bandwidth 15–16, 83, 88; international 88
Belgrade Embassy bombing 145–6
biotechnology 9, 107, 109, 113, 119
Boyle, James 3, 60, 68, 69, 73
broadband 12–17, 90, 97, 98–9, 105; (convergence) 5, 23, 83, 85, 91, 94
Buzan, Barry 155

cable TV 4, 12, 83, 84, 89, 90, 93, 97; penetration and subscribers 91; stations 91
Chen, Shui-bian 106, 108, 118, 152
Chen Xiaoning 95
China: Central Television (CCTV) 90, 92, 151; Education and Research Network (CERNET) 88; Internet Exchange (CNIX) 88; Netcom 95; Netcom Jitong 15; Railcom 15, 99; Unicom 15, 19; Telecom 13, 46, 48, 85, 88, 89, 93, 94, 133
Chinese: Academic Network (CANET) 131,32; Academy of Sciences (CAS) 9, 98, 133, 149; Domain Name Consortium 135; People's Political Consultative Conference 66, 155
Chinese Communist Party (CCP) 2, 4, 21, 35, 65, 66, 83, 84, 85, 90, 92, 94, 96, 98, 140–1, 151; Central Policy Research Office 141; Fifteenth Congress 143; Propaganda Dept. 89, 90, 93, 98, 99; Sixteenth Congress 97
Cisco Systems 69, 103, 116
Clinton, Bill 30, 58, 130
CNN 66, 134
CNNews.com 134
CNNIC 20, 31, 34–5, 128, 132–3
convergence (technological) 21–3, 83; tri-net (*sanwang ronghe*) 84, 5, 9, 93, 95–8
crime: computer 147, 153; cyber 59; arms trafficking, drugs, money laundering, organised 152–4; *see also* pornography, terrorism
Cyberport 105, 107, 110, 114, 122

Dalai Lama 150
databases 69, 154
Democratic Progressive Party (DPP – Taiwan) 106, 116, 118, 152
democratisation 3, 30, 68, 73, 131
Deng Xiaoping 9, 73, 97, 141
Department of Commerce (USA) 5, 130, 137
Department of Defense (USA) 148
digital divide 3, 9, 10, 20, 30, 38 42, 99; gap with USA 147

Eastern Communications Co. (Eastcom) 134
e-commerce 2, 10, 20–1, 25, 35, 59, 130
education 10, 35, 44, 46, 50–2, 61, 72, 104, 106, 119
employment 38–40
encryption 60–1, 71
Enterprise Online 12, 18, 25
EP-3 incident 70, 145, 152, 153

176 *Index*

Falungong 65, 69, 71, 145–6, 151
Family Online 12, 25
Fang Yihong 93
Fangzhou-1 processor 18
fibre-optic cable (national grid) 13–15, 66, 83–4, 90, 93
Five Year Plan: Ninth (1996–2000) 142, 147; Tenth (2001–5) 10, 11, 17, 35, 66, 83, 84, 96, 148, 155
Formosa Plastics Group 116–17
Foucault, Michel 3, 60, 68

G7/8 2, 10
Gates, Bill 107, 150
General Electric Information Services 116
Global Information Infrastructure (GII) 140, 147
go west policy 35, 46, 117
Goh Chok Tong 102
Golden Shield 69, 150
Gore, Al 140, 147
Government Online 12, 25, 35, 37, 58, 65, 72
Green Silicon Island 106–9

hacking 69, 139, 145–6, 152–3; *see also* EP-3 incident
Hong Kong 5, 24; at CPPCC 155; domain names 134, 135; in Greater China 102–23; Industrial Technology Centre 107, 13; and Shanghai Banking Corporation 114; Silicon Valley Association 114; Stock Exchange 114; Telecom 150; University of Science and Technology 119
Hsinchu Science-Based Industrial Park 105, 108, 115
Hu Jintao 21
Huang Qi 65

information warfare 139–40; cyber warfare 140, 45, 52; people's war 140, 143; Sun Zi 139; trapdoors 139; Trojan horses 139; *see also* viruses
Information Technology Development Leading Group (ITDLG) 21
Informatisation of the National Economy Programme 9, 10
Intel 18, 103; Pentium III serial numbers 142
intellectual property rights 12, 61
Internet 20–1, 23, 24, 25, 30, 83, 90, 98–9; Assigned Numbers Authority (IANA) 129; cafes 64, 66–7, 71, 149; clinics 146; commercialisation of 35; Content Providers (ICPs) 60–1, 64, 69, 73, 84; Corporation for Assigned Names and Numbers (ICANN) 130–9; clientele base 35–7, 40, 42, 72; Engineering Task Force (IETF) 129, 135; Great Wall 143; number of users 31–4, 83; police; 146; regional disparities 42–53, (maps) 41, 43, (charts) 42, 44; Service Providers (ISPs) 24, 48, 60–1, 67, 73, 84, 153

Japan 140; Bank of, attacked 145
Jiang, Simon 119
Jiang Zemin 1, 9, 11, 38, 58, 96, 104, 140, 143, 147, 151, 155

Keohane, Robert 134
Kuomintang (KMT) 116

language 35, 59, 72, 84, 131; Chinese 120, 134–6
law: domestic 153; international 152; ISPs 153
Lee Teng-hui 115, 145
Legend Computers 68, 149, 150
Lessig, Lawrence 3, 59, 73
Levin, Gerald 151
Li Ka-Shing 113–14
Li, Richard 107, 110, 114
Li, Zibin 104, 106, 121
licences, for Internet services 63–5, 71
Lin Hai 65, 146
Linux, Red Flag 38
Lu, Annette 117

Maya 134
Measures for the Administration of Radio and Television (1997) 93
Mi, Zhenyu 155
Miao, Matthew 116
Microsoft 19, 71, 107, 108, 120; Windows 38, 52, 142, 150
Military Affairs Salon (*People's Liberation Army Daily*) 142
military doctrine: strategic boundary 143; strong technological army 143, 144; *see also* information warfare
Mills, David 128
Ministry of Economic Affairs (Taiwan) 115
Ministry of Education 132
Ministry of Electronic Industry (MEI) 85, 132
Ministry of Information Industry (MII) 11,

22–3, 25, 31, 83, 84–5, 88–9, 92, 94–5, 141, 150, 155
Ministry of National Defence (Taiwan) 144, 150, 152
Ministry of Posts and Telecommunications 13, 22, 23, 85, 132
Ministry of Public Security 69
Ministry of Radio Film and Television 22, 89
Ministry of Railways 46, 98
Ministry of Science and Technology 118, 119
Mitac-Synnex Group 116
mobile communications 12, 17, 19
Mockapetris, Paul 128
Monte Jade 105
Motorola 19, 48
Multilingual Internet Names Consortium (MINC) 131
Murdoch, James 151
Murdoch, Rupert 24, 147, 150–1

Nankang 106; Software Park 115
National Computer Networking Facility of China 132
National Information Infrastructure (NII) 25, 105
National Informatization Leading Group 95
National People's Congress (NPC) 66, 84
National Science Council (Taiwan) 105; White Paper on Science and Technology (1997) 105, 107, 115
National Science Foundation (USA) 3, 129, 132
national team 149–50
nationalism 145; nation-building 122, 156; patriotism 18; unity of nationalities 62, 70
NATO 145–6
natural economic territories 102
Netcom 93
network access points (NAPs) 88
Network Civilization Project Organizing Committee 155
Network Solutions Inc. (NSI) 129
New Culture Forum 65
new economy 9, 12, 24, 30, 35
New Middle Way 106
news: provision of 62–3; electronic newspapers 142
News Corporation 151
Nokia 19

Nortel Networks 69
Nye, Joseph 139

Office for Information Industry 95
Open Source movement 129
Oracle 120

People's Daily 58, 63, 68, 70; Strong State Forum 70–1
people's war 143
People's Liberation Army (PLA) 12, 148; Academy of Military Sciences 155; Airforce Academy 145; *PLA News* 142
personnel 48, 50–2, 64, 69, 106 148; military 144; Taiwan 117
Phoenix Satellite Television 24
Plan to Develop a Knowledge-based Economy in Taiwan (2000) 106
pornography 62, 66, 69, 152
Postel, Jonathan 128–9, 131
Public Security Bureau 64, 66, 71

Qian, Tianbei 132, 133
Qiao, Liang 145
Qiao, Shi 119
Qinghua University, Beijing 66
Qiu, J. Linchuan 61, 68

Railcom 46
Reagan administration 148
regions and regional disparities 20, 35, 40, (maps) 41, 43, (charts) 42, 44, 50
regulation (and legal framework), 24, 59–65, 73, 85, 96, 97, 99
religion 62; *see also* Falungong
Revolution in Military Affairs (RMA) 139, 143, 155

SafeWeb 72, 154
satellite broadcasting 24, 46, 90
Security China 2000 69
September 11 incident 6, 59, 71–2, 153–5
Shanghai 15, 19, 20, 40, 48, 88, 97, 117, 118, 134, 147, 149; Municipal Government 98, 119; Six 154
Shenzhen 40, 103–4, 108, 109, 118, 120–2, 150
Siemens 19
Silicon Bridge 122–3
Silicon (plus) Coalition 116–17, 122
Silicon Valley 5, 48, 102–23
siliconisation 5, 103–4, 107, 116, 122, 140

178 *Index*

Sina.com 63, 69, 70
Singapore 64, 102, 109; in Greater China 119–20; domain names 135
software: availability 52; development 19; houses 19; for censorship 60, 69
Sohu 62
Stan Shih 108, 116–17
Star TV 24
State Administration of Radio, Film and Television (SARFT) 4, 83, 84, 85, 89–95
State Council 21–2, 63, 67, 89, 92–3, 94; Information Management Commission (SCIMC) 21; Informatization Leading Group 23; Department of Science and Technology 104; Development Planning Commission (SDPC) 18, 95; High Technology Research and Development Plan 148; Strike Hard campaign 154

Taiwan: in Greater China 5, 102–23; domain names 130, 135, 140; Strait Crisis 143, 145; information warfare 144–5, 152
Tan, Tin Wee 131, 135
TD-SCDMA 19
Telecommunications and Information Association (USA) 130
Telecommunications Regulations 94
Telecommunications Law 23, 25
telephone networks 16, 17, 24, 35, 44, 46, 48, 83, 85; subscribers 90
terrorism 6, 152, 154
Third Generation (3G) mobile communications 19
Third Wave 8
Third Way 106
Three Represents campaign 2, 143
Three Worlds theory 6, 141
Tian Congming 93
Tiananmen Incident 65, 70
Top Level Domains (TLDs) 128; country code 129, 130, 132; generic 129, 132
Triangle Boy 72

Tsang, Donald 104, 110
Tung Chee Hwa 104, 107, 110, 114

United Nations 2, 38, 130; Charter 152; Security Council 154
United Nations Development Programme (UNDP) 10, 50–1
United States: data sharing 154; DNS 5, 127–8, 130, 134, 136; Falungong 66; hegemony 71; information warfare 140, 142, 143, 145–6; new economy 9; *see also* Triangle Boy

VeriSign Inc. 128, 130, 134–6
VIP Reference (Dacankao) 65, 146
Voice of America 72
viruses 67, 139, 142–3, 153

Walton, Greg 149–50
Wang, Xianghui 145
Wang, Xiaoquan 93, 94
Websites: blocking 66; content 35, 37, 44, 84; licenses 63–4; number and location 20, 31, 40; Sacred Sovereignty 146
World Trade Organisation (WTO) 23–5, 46, 61, 84, 89, 95–6, 120, 134; China–US Agreement 150
Wu Jichuan 11, 136, 147

Xinhua News Agency 63
Xinjiang separatist movement 154
Xu, Guangchun 89, 92, 94, 151
Xu, Wenbo 155

Yahoo! 107, 109
Yang, Jerry 107, 109
Yang, Yixian 155

Zeng Peiyan 95
Zengcheng 108
Zhang Chunjiang 22, 155
Zheng, Cindy 132, 133
Zhu Rongji 8, 9, 95, 98, 104, 107
Zou Jiahua 21